Ninox boobook

The other use for species names in life (sciences)

A Science Diction by
Jörn Glökler

Die Deutsche Nationalbibliothek verzeichnet diese Publikation in der Deutschen Nationalbibliografie; detaillierte bibliografische Daten sind im Internet über dnb.d-nb.de abrufbar.

ISBN 978-3-8391-8101-0
Herstellung und Verlag / printed and published by:
Books on Demand GmbH, Norderstedt, Germany

To Antje and Yorik

with many thanks to my colleagues who were very helpful for (partially) involuntary inspiration and very voluntary critique.

Also special thanks to Sean O'Keefe and James Adjaye for watching my language.

This book was made possible by freely available databases and software.

Software

Dictionary database and book layout
OpenOffice (http://www.openoffice.org)
Vector graphic design
inkscape (http://www.inkscape.org)
Raster graphics
The GIMP (http://www.gimp.org)
Operation system
Ubuntu Linux (http://www.ubuntu.com)

Taxonomy databases

ITIS (http://www.itis.gov)
and NCBI (http://www.ncbi.nlm.nih.gov/Taxonomy/)

Graphic templates

Wikimedia Commons (http://commons.wikimedia.org)

Preface

Just as Eskimos are supposed to have an extensive vocabulary for snow including the yellow variety, scientists should also have a similar range of vocabulary especially when it comes to experiments.

There are those experiments that are always dead certain to work, but simply refuse to work in your hands. There are other experiments that get delayed because you're waiting for your shipment despite repeated reassurance from the supplier that it should arrive any time soon. Then there are experiments you perform out of schedule while still waiting for your samples to arrive, and those experiments that other colleagues already get published just when your long awaited samples finally get delivered.

These and other related issues in the daily routine of a life scientist deserve a name so that it can be used instead of less desirable and creative four letter words that may normally spring to mind. Following the example of "The meaning of Liff" written by the late Douglas Adams and John Lloyd, I have chosen another vast source of vocabulary that is more appropriate for life science than city names. This is a great opportunity to put another unexpected use to the otherwise less uplifting collection of words just waiting to be exploited: the species names in taxonomy.

This slightly "other" dictionary will breathe new life into some otherwise dusted and forgotten species names and give them the attention they deserve in the scientific community.

A

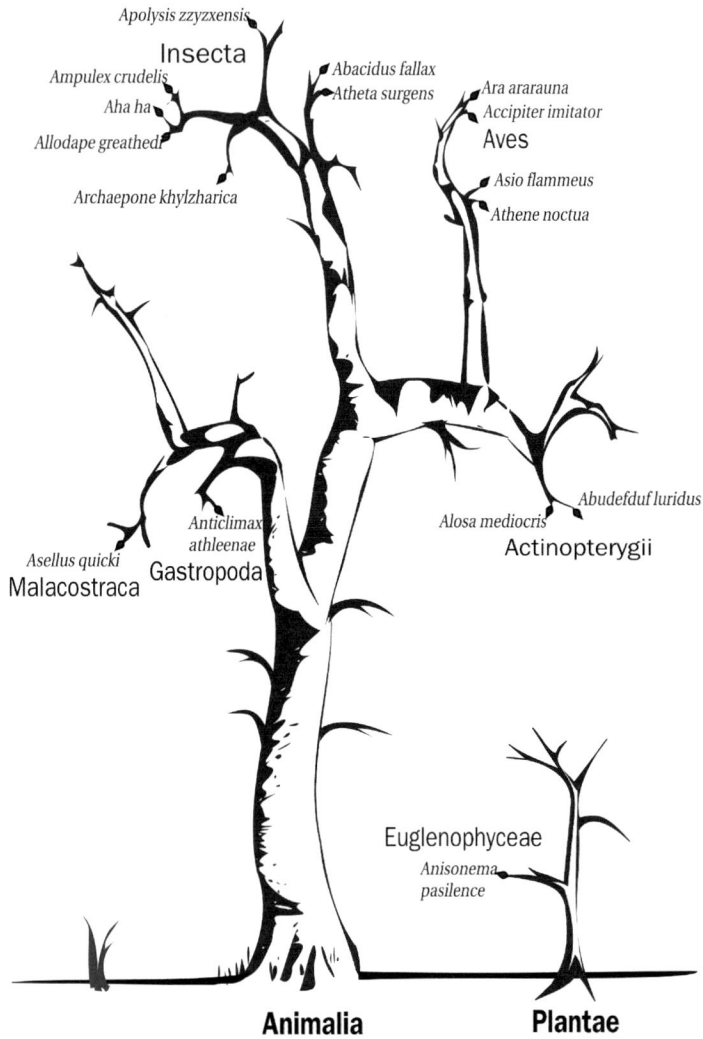

Abacidus fallax

the inevitable battered lab calculator reminiscent of the electronic dark ages.

Abudefduf luridus

the lecture rendered unintelligible due to a strong foreign accent, speech impediment / forgotten dentures.

Accipiter imitator

the deceptive label on glass bottle indicating the wrong content. For some reason such labels are mostly found on commonly used stock solutions.

Aha ha

the prank played by colleagues by swapping your ELISA plate with a fake one. Some carefully arranged pattern using small amounts of enzyme will usually do the trick.

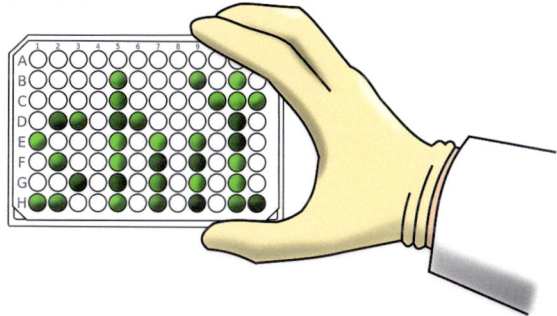

Allodape greatheadi

biggest scientific discovery ever made. Each professor claims at least one allodape his own.

Alosa mediocris

the second authorship position marked as "··· contributed equally".

Ampulex crudelis

the special ingredient that makes your experiments reproducible but remains elusive. Ampulex is mostly found in bidest water, glassware and on the hands of certain technitians. As long as things are running smoothly: don't change your institute or team.

Anisonema pasilence

the uneasy feeling caused by the dead silence after you finished your presentation without any questions.

Anticlimax athleenae

the discovery that your discovery has been discovered before.

Apolysis zzyzxensis

the rolled up rejected manuscript now most useful as a fly swatter.

Ara ararauna

the distinct noise produced by a vortexer with worn rubber feet on a stony lab bench.

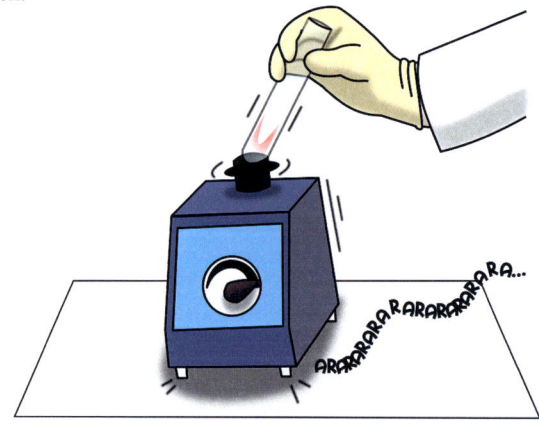

Archaeopone kzylzharica

the scientific term you don't know how to spell or even to pronounce correctly. It is also the best candidate to become your next histiophryne (q.v.).

Asellus quicki

the annual spending frenzy. Mysteriously, missing funds suddenly surface on the very last day in the year they can be spent. Best opportunity for sales representatives to sell their otherwise shelf-warming usnea vainioi (q.v.).

Asio flammeus domingensis

the fire alarm coinciding with the final stage of your long term experiment.

Asio flammeus sandwichensis

the charred remains of what caused the fire alarm that's left in the kitchens' microwave.

Athene noctua impasta

the problem of getting night owls and early birds to coordinate their experiments in the lab.

Atheta surgens

the cool and misty foam emerging from a sink caused by a mix of water, detergent, and dry ice mysteriously coinciding with enzyme deliveries.

B

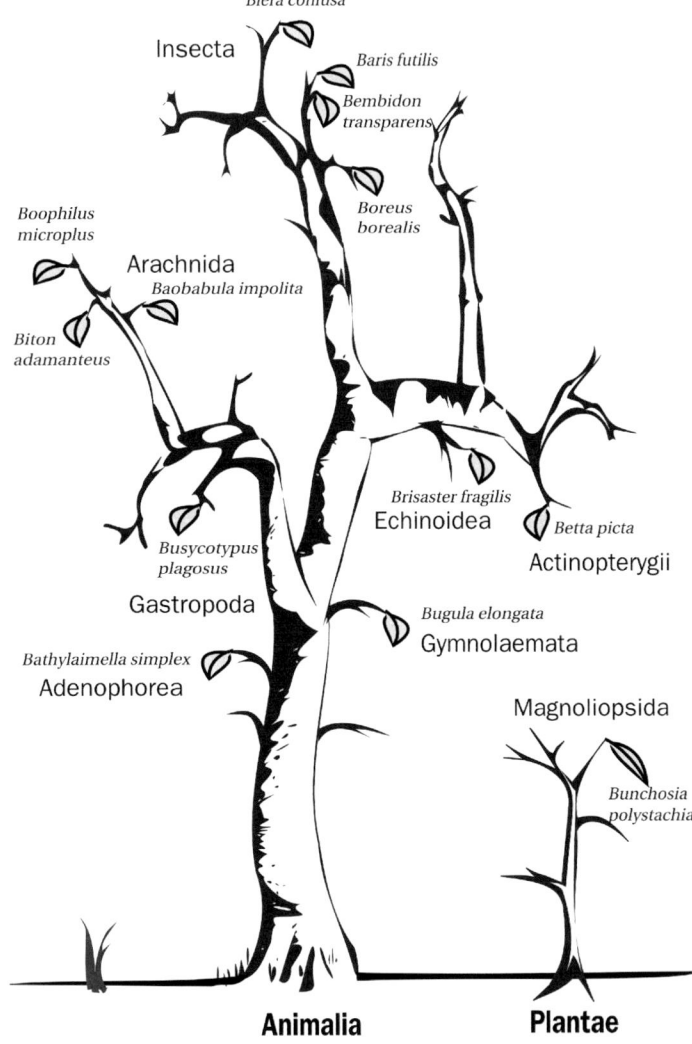

Blera confusa

Insecta

Baris futilis

Bembidon transparens

Boreus borealis

Boophilus microplus

Arachnida
Baobabula impolita

Biton adamanteus

Brisaster fragilis
Echinoidea

Betta picta

Actinopterygii

Busycotypus plagosus

Gastropoda

Bugula elongata
Gymnolaemata

Bathylaimella simplex
Adenophorea

Magnoliopsida

Bunchosia polystachia

Animalia

Plantae

Baobabula impolita
the background murmur during a plenary lecture. Severe baobabula includes finger pointing and giggling.

Baris futilis
the identical experiment that returned positive results yesterday which is coming up negative today.

Bathylaimella simplex
one of those improvised styrofoam floats found by default in all water incubators

Bembidion transparens
the acrylic radiation protection screen. It is especially designed in a bulky form so that you can only access your samples with either very long arms or by turning it aside.

Betta picta
the perfect shot that's playing hard to get as the centre piece of your intended publication. Each time you try some other part is either blurred or distorted.

Biton adamanteus adamanteus

the block of salt you can't pry a pinch from no matter how much you bang the container against the wall. Also a perfect tool to transform spatulae into works of Uri Geller.

Blera confusa

the state of mind when lab vocabulary starts to leak into private life such as: "could you get me the ice cream from the -20?".

Boophilus microplus

one drop too many in any given pH titration experiment.

Boreus borealis

the state of mind quickly induced during the annual safety lecture shortly before switching to autopilot mode.

Brisaster fragilis

the pet theory of a senior scientist. Many indicator minor minor (q.v.) have helped to defend it against the common understanding of the scientific community.

Bugula elongata
the colleague that keeps nagging you all day about every single detail in a simple protocol.

Bunchosia polystachia
any number of permanent markers in your lab coat that turn out to have run dry.

Busycotypus plagosus
the colleague who at first is very eager to collaborate in a project, then claims to be far too occupied to contribute, but finally demands to be on the paper.

C

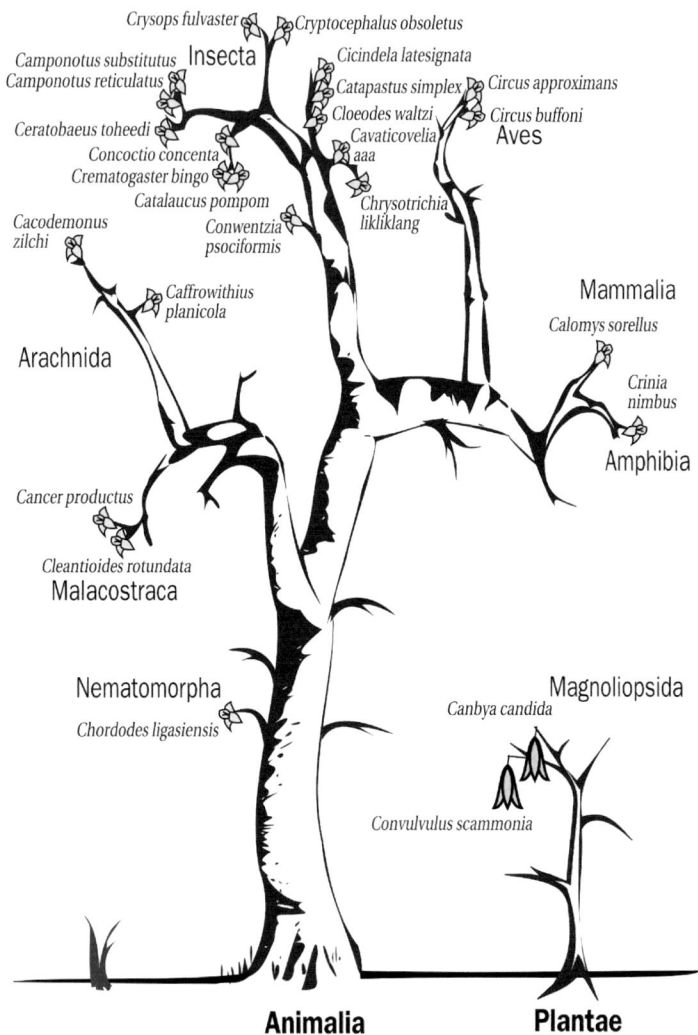

Cacodemonius zilchi

the ominous entity that dwells in all labware and feasts solely on RNA.

Caffrowithius planicola

the visitors that reliably show up during your coffee break.

Calomys sorellus

the sore developing at the tip of your thumb through working with too much plasticware.

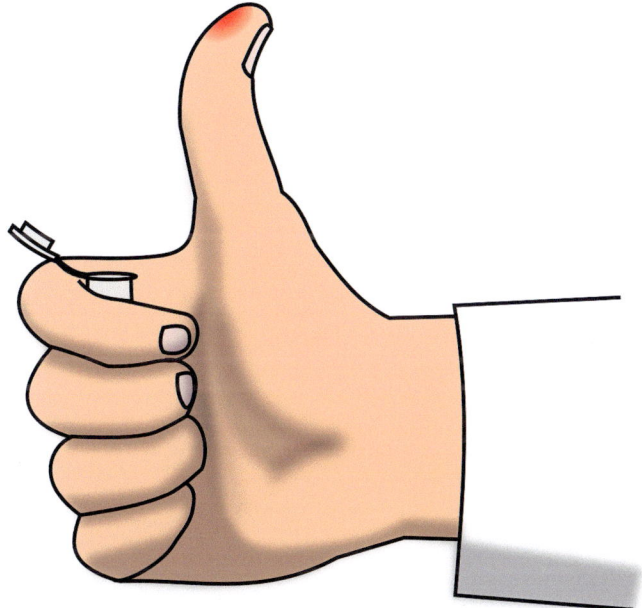

Campanula bononiensis

the typical medicinal research paper claiming relevance to the whole population based on a study involving two related individuals living next door to each other. Perfectly fine until the information surfaces that a pharmaceutical company funded the research.

Camponotus reticulatus fullaway

The problem with discontinued and highly coveted parts for certain old devices. Especially when the stock of replacement part runs out and you will have to resort to camponotus substitutus multiplis (q.v.).

Camponotus substitutus multiplis

the heavily improvised repair of unavailable parts for discontinued devices. Especially challenging if it came from a defunct company formerly situated in a country now at civil war.

Canbya candida

the institute's vending machine that allows you to finish your late night experiments.

Cancer productus

yet another gene therapy gone awry.

Canis lupus familiaris

the head start in scientific career gained by mere inheritance.

Catapastus simplex

the results that other colleagues get published just when your long awaited samples finally get delivered for the same intended experiment.

Cataulacus pompom

the submerged round and fluffy stuff that can be found in once sterile buffers.

Cavaticovelia aaa

first and most important experiment in the morning right after a cup of coffee, a good chat, and weeding out your incoming mailbox.

Ceratobaeus toheedi

the apparently silly advice by the senior scientist or technician. Ceratobaeus often turns out to be quite useful after you have found it out for yourself.

Chordodes ligasiensis
the most potent ligation spell known to science on this side of reality. It also helps keeping warthogs from your door.

Chrysops fulvaster
the colleague that shows up first despite being the last to leave yesterday's lab party.

Chrysotrichia likliklang
the sound of the microscope lens passing through the slide.

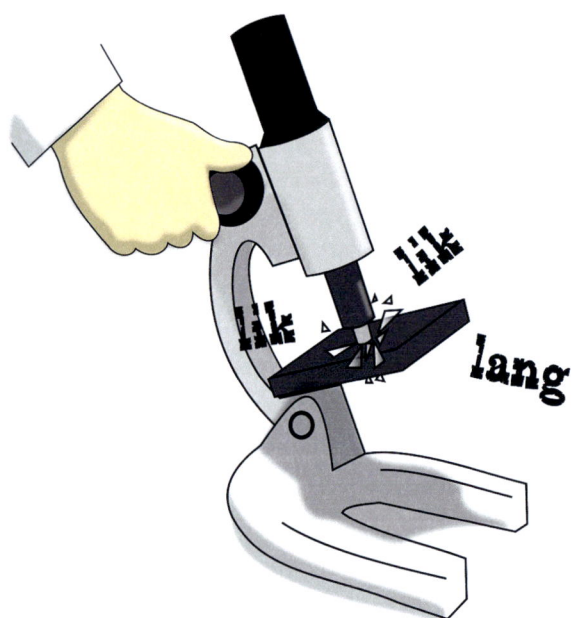

Cicindela latesignata obliviosa
just one minute past the grant application online submission deadline.

Circus approximans

when you know that the next lab seminar presentation will be yours to hold.

Circus buffoni

the guy asking silly questions during your lab seminar. Often it's the supervisor claiming to do you a 'favour', or a nasty colleague that couldn't care less about his own turn.

Cleantioides rotundata

the dirty spaces found underneath and sometimes inside bulky devices where nobody cared to clean for ages. Cleantioides are a valuable repository for all microorganisms that have been handled in the lab over the years.

Cloeodes waltzi

the experimental steps taken to find the right approach to a novel scientific problem without any further information. One step ahead, some sidetracking, one step back, and with some luck you'll end up at square one.

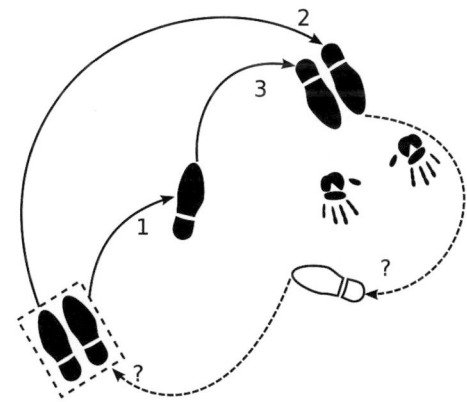

Concoctio concenta

the long-term chemical evolution experiment secretly taking place in the huge organic waste container under the hood. Unfortunately, never analysed before disposal.

Convolvulus scammonia

the unsupported findings that lead to snake oil-like claims in life science. Convolvulus helps acquiring huge amounts of venture capital to build up biotech startup companies. When all money is spent after 3 years, a new company is founded based on an even more preposterous convolvulus.

Conwentzia psociformis

a typical group of smokers gathering outside the congress building.

Copidognathus pasticus

the digital skill to quickly generate results, thesis, papers, grant applications, and career problems when finally caught.

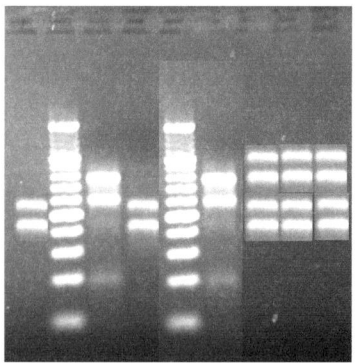

Cora dualis

the kitchenware listed in a labware catalogue. Best buy for aroma extractors if for some reason caffeine-rich solutions need to be prepared on a daily basis.

Costora luxata

journals belonging to the access of evil.

Crangon handi

the brass knuckles equivalent of molecular biology. A perfect tool to vent one's anger by closing PCR strips. Hence the frequent expression "I'm so mad I could do some PCRs right now!"

Craponius inaequalis

the results obtained from the same experiment performed by different people.

Crematogaster bingo

when all quadruplicate samples yield the same value. This indicates that the experiment is either highly reproducible or more likely, that the instrument is broken.

```
##BLOCKS= 2
Plate:    22 4rd scFv 1 01. Mrz PlateFormatEndpoint  AbsorbanceRaw   405   1  12  96   1   8 None
          Temperature( C)   1        2        3      4   5   6   7   8   9  10  11  12
19.40     A      0.025    0.028    0.027    0.028     0   0   0   0   0   0   0   0
          B      0.045    0.045    0.047    0.048     0   0   0   0   0   0   0   0
          C      0.162    0.162    0.162    0.162     0   0   0   0   0   0   0   0
          D      0.071    0.07     0.083    0.07      0   0   0   0   0   0   0   0
          E      0.003    0.043    0.105    0.102     0   0   0   0   0   0   0   0
          F      0.025    0.026    0.022    0.028     0   0   0   0   0   0   0   0
          G      0.032    0.033    0.03     0.031     0   0   0   0   0   0   0   0
          H      0.01     0.011    0.009    0.018     0   0   0   0   0   0   0   0

~End
```

Yes!

Crinia nimbus

the status lost by the first experiment in your career that's not dead certain or even rigged to work. Most people lose their crinia during their bachelor thesis. From then on experimental failure is promoted from exception to rule.

Cryptocephalus obsoletus

the unexpected extra feature of some lab instruments. The pH meter that also works as a perfect random number generator, or the tabletop centrifuge that could pass as a supersonic jet engine.

D

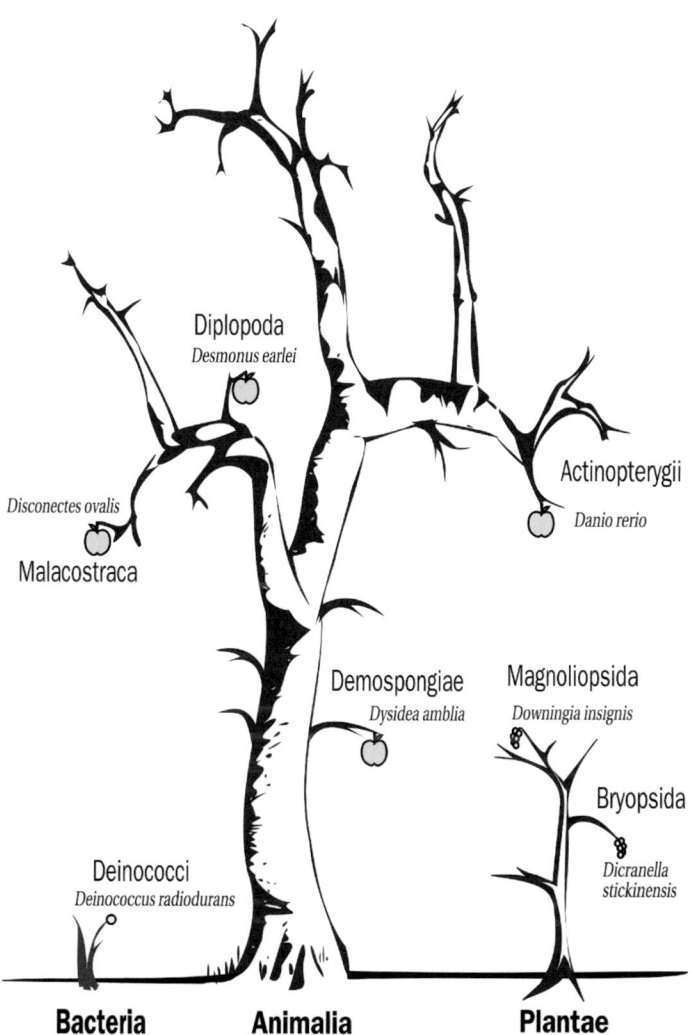

Diplopoda
Desmonus earlei

Actinopterygii

Disconectes ovalis

Danio rerio

Malacostraca

Demospongiae
Dysidea amblia

Magnoliopsida
Downingia insignis

Bryopsida

Deinococci
Deinococcus radiodurans

Dicranella stickinensis

Bacteria   **Animalia**   **Plantae**

Danio rerio

the kind of guy that reuses his latex gloves until they start to leak.

Deinococcus radiodurans

a long term social experiment. Two students with mutually exclusive preference for background radio music are put into one lab during their PhD studies. The winner is the one to finish the thesis first.

Desmonus earlei mancus

the most interesting sample which is also prone to expire first.

Dicranella stickinensis

the pretentious sticker found on a colleague's lap top cover or car reading "if it ain't in Medline it ain't published".

If it ain't MEDLINE
It ain't published!

Disconectes ovalis

when you try to get some important device working again only to realise that it was unplugged all the time.

Downingia insignis

the long way down from the initially targeted impact factor journal to the finally accepted toilet paper.

Dysidea amblia

seeming to know what's the cause for the unexplained phenomenon, but each time you try to get close to a solution it manages to escape.

E

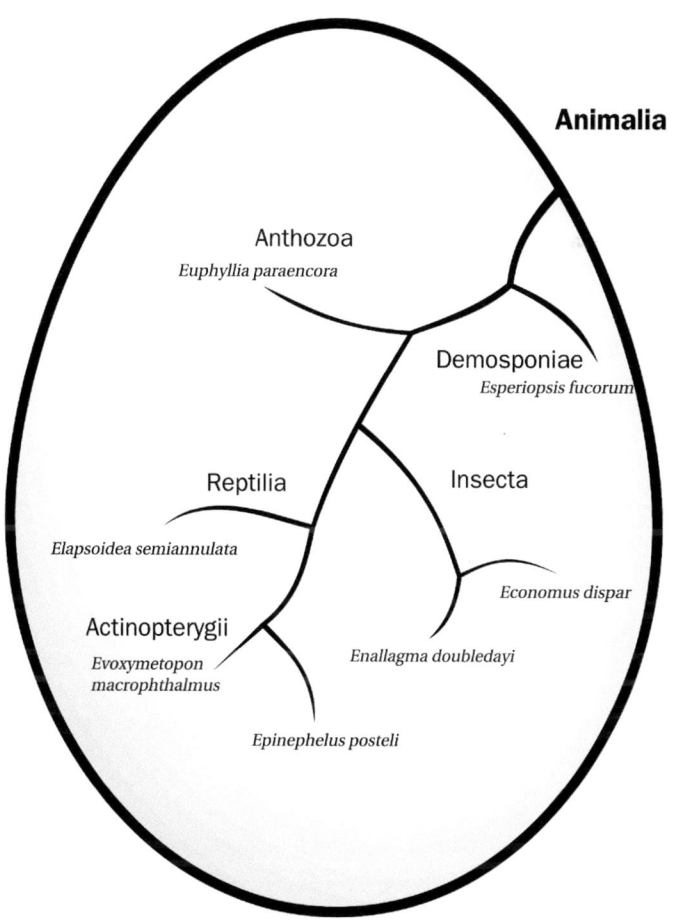

Ecnomus dispar

the subtle but painful financial difference between the original project application and the final grant.

Elapsoidea semiannulata moebiusi

the recurring writer's block when the six-monthly project report is due.

Enallagma doubledayi

the experiment you couldn't finish late at night no matter how hard you tried.

Epinephelus posteli

the frayed corners of a frequently recycled poster.

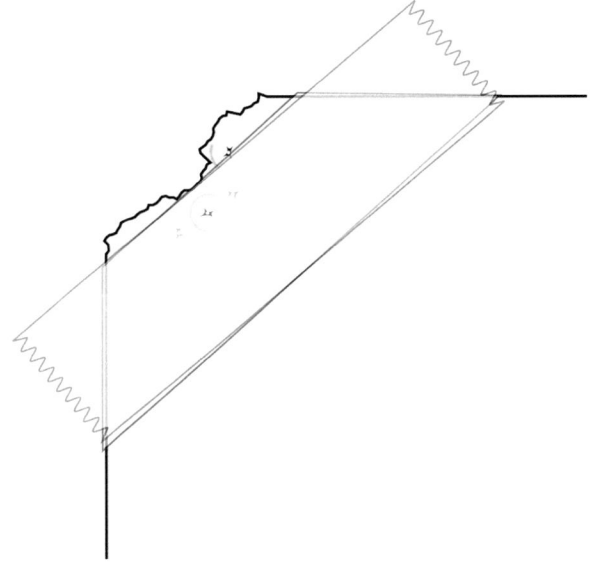

Esperiopsis fucorum
the controls that spring to mind after you've done the experiment.

Euphyllia paraencora
the sudden realisation that you will have to start all over with your project after you noticed that all respective samples were claimed in the clean up of the fridge during your 4 week vacation. Lesson learned: never go on holidays unless you really have to.

Evoxymetopon macrophthalmus
the spontaneously invented scientific term used as a defence in front of a high-profile scientific audience that nobody dares to challenge.

F

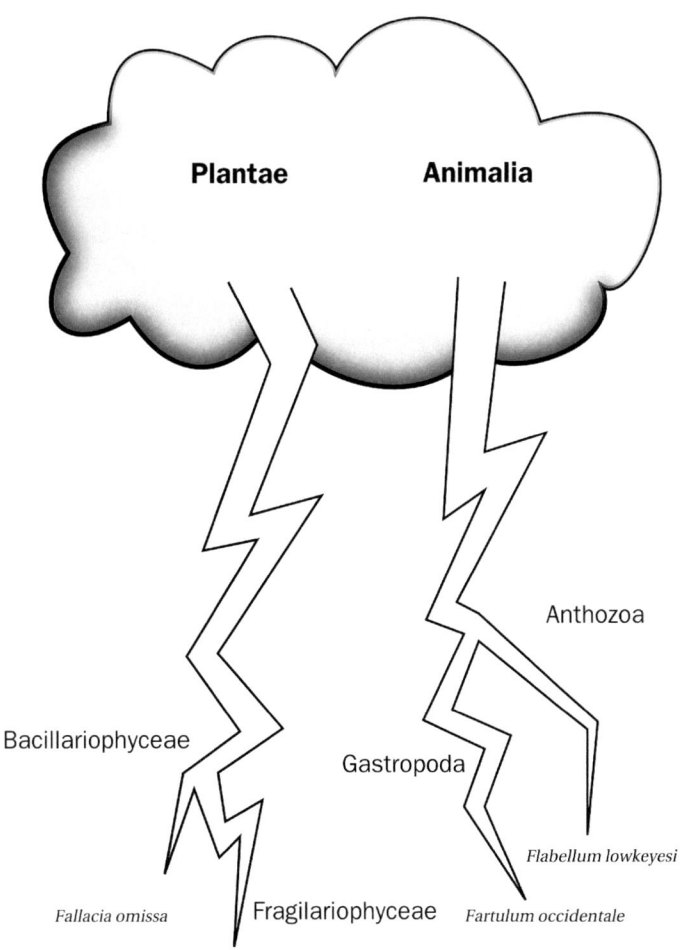

Fallacia omissa

the somewhat confusing data that wouldn't fit into the presented theoretical model and doesn't make it into the final report.

Fartulum occidentale

the embarrassing mishap difficult to cover up in time with a dose of mercaptoethanol.

Flabellum lowkeyesi

the difficulty not blurting out your findings before the paper or patent application is submitted.

Fragilaria lata

the shipment that turned out to be much less robust than expected.

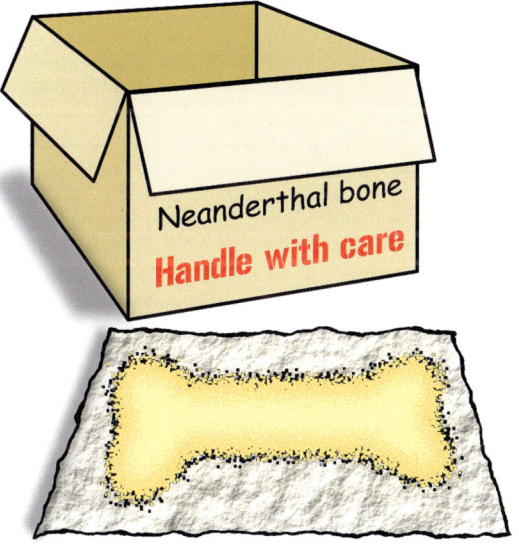

G

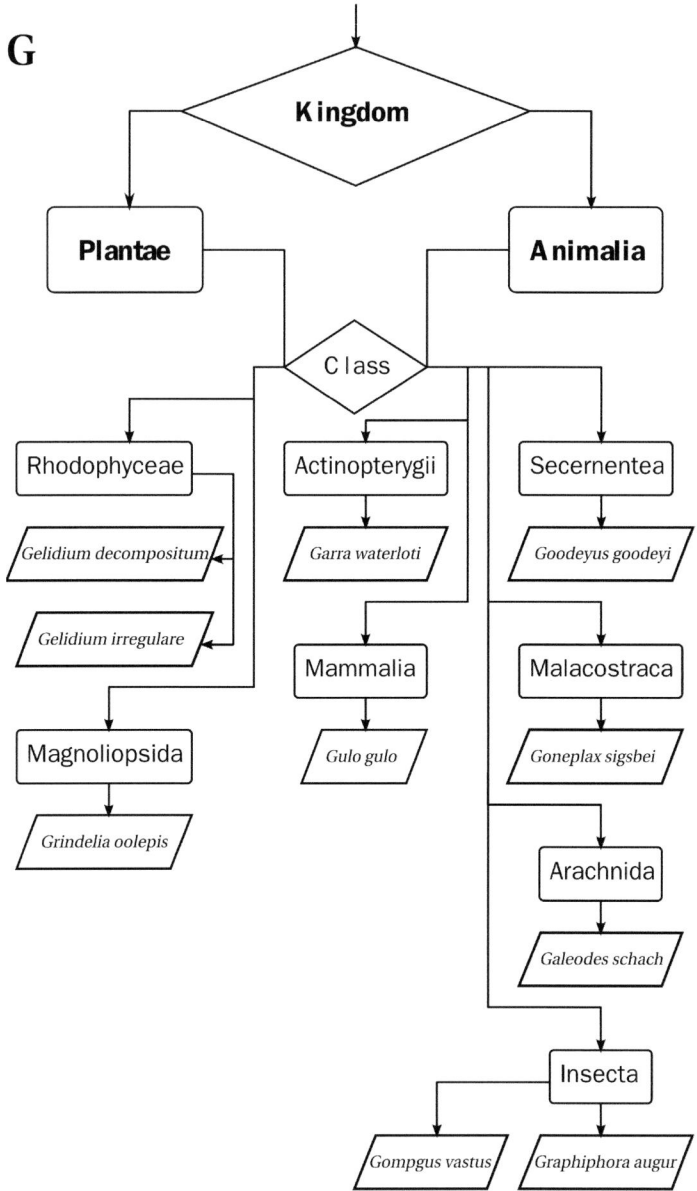

Galactia smallii

the situation when you meet somebody on a congress that you least expect. Quite often the same person you're least happy to meet and being exposed to during the entire congress anyway.

Galeodes schach

the shape of a colony, blot, microarray spot, or coffee stain that look entirely like something else. Time to take some days off if this happens more frequently.

Garra waterloti

the substance that wouldn't dissolve as well as expected. Eventually resulting in 0.2 x stock solutions.

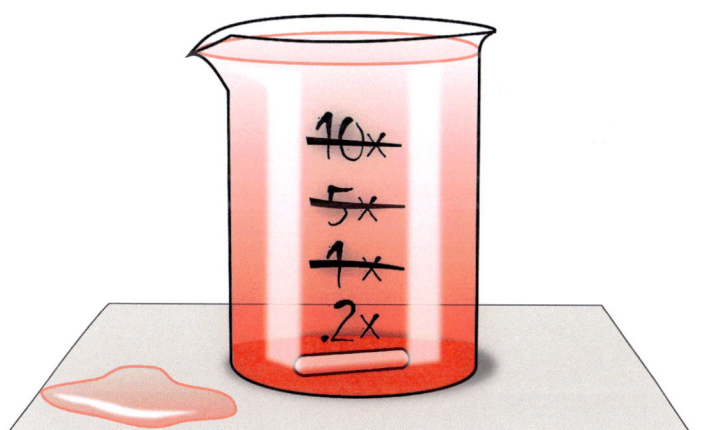

Gelidium decompositum

the bits of gel that get stuck between the comb's teeth.

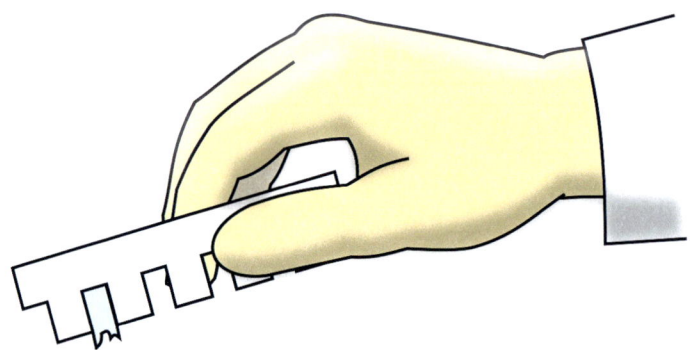

Gelidium irregulare

the undulating bands resulting from a leaky electrophoresis chamber.

Gomphus vastus

the texture of stuff at the very bottom of the autoclave bin after final treatment. It is proposed that the primordial soup from which all life descended was gomphus, too.

Goneplax sigsbei

the minuscule speck left of the precious sample after your sneeze blew it off the analytic spoon.

Goodeyus goodeyi

the annoyingly cheerful guy in your lab that isn't fazed by anything.

Graphiphora augur

the decent share of the sample that is sacrificed in an ancient ritual to the oracle for analysis. Unfortunately as with any advice obtained by augurs, proper interpretation is the tricky part. To make sure that the interpretation is correct another oracle is to be consulted, thus doubling the amount of offered graphiphora.

Grindelia oolepis

the parts of your equipment that don't belong into the homogeniser.

Gulo gulo gulo

the hidden drinky drinky motion targeted to fellow scientists. Gulo is very useful during boring and lengthy lab meetings to signal that an after work beer is to be arranged once the supervisor is gone.

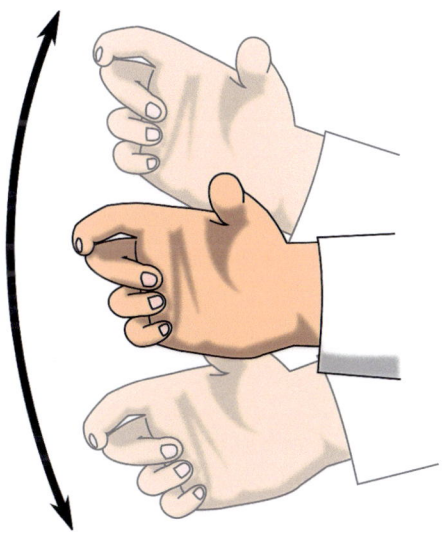

H

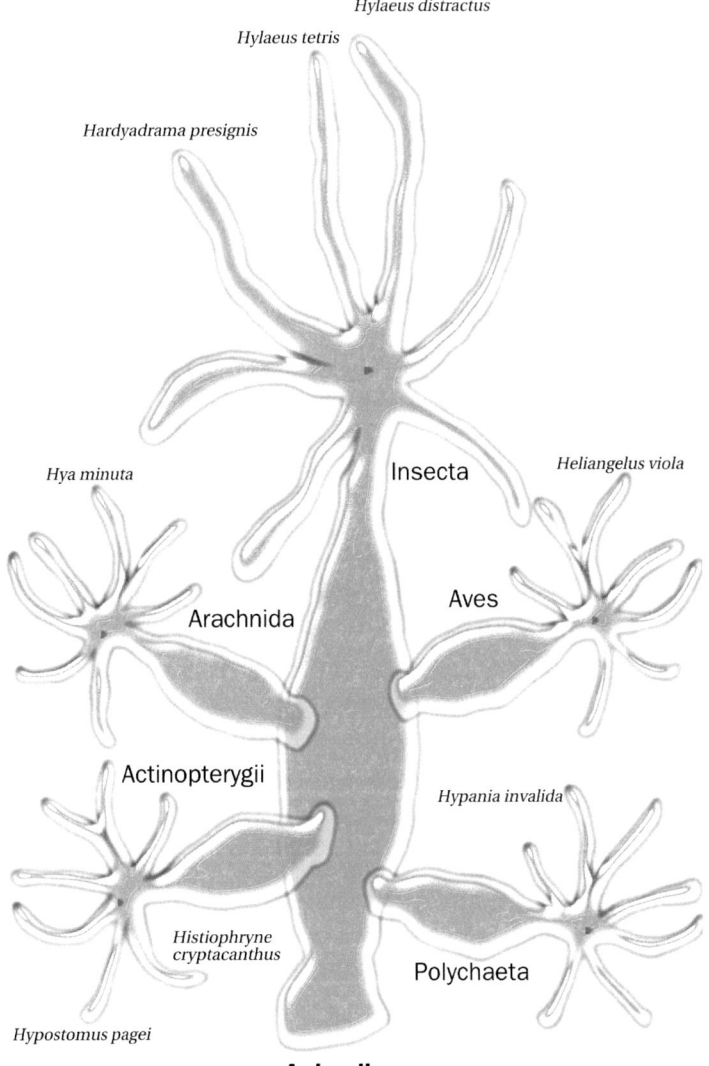

Hylaeus distractus

Hylaeus tetris

Hardyadrama presignis

Insecta

Hya minuta

Heliangelus viola

Arachnida

Aves

Actinopterygii

Hypania invalida

Histiophryne cryptacanthus

Polychaeta

Hypostomus pagei

Animalia

Hardyadrama presignis
the inevitable argument that ensues with Kanaloa manoa (q.v.) over participation at a Kanaloa kahoolawensis (q.v.).

Heliangelus viola
the strong response to a failed experiment that instantly leads you to supect your fellow scientist being a member of a biker gang.

Histiophryne cryptacanthus
the ultra-safe password that you just can't seem to remember.

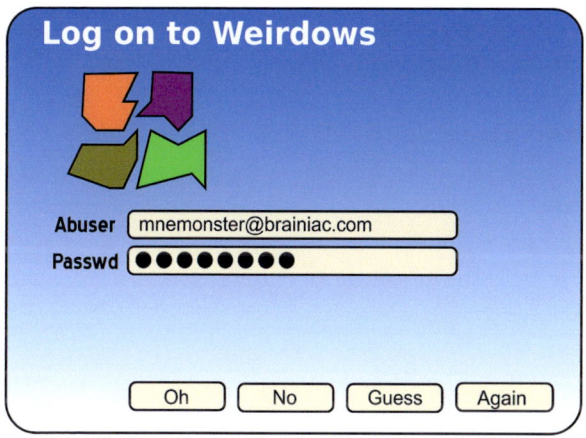

Homalota flexibilis
the certain advantage when you're into bioinformatics.

Hya minuta
the unexpected brief conversation in the corridor that can well last more than an hour. Hya minuta is most likely interrupted by the urgent need to visit the ceramics department.

Hylaeus distractus

any welcome activity like a Hylea tetris (q.v.) to pass the time without doing what needs to be done.

Hylaeus tetris

the simple online game burning up more time in the lab than cigarette breaks.

Hypania invalida

the next big thing in life science. All your colleagues flock to it only to realise that it doesn't live up to its initial promise and are also too late for any of the anticipated high impact papers.

Hypostomus pagei

the time point you notice that the electrophoresis is running on reverse polarisation.

I

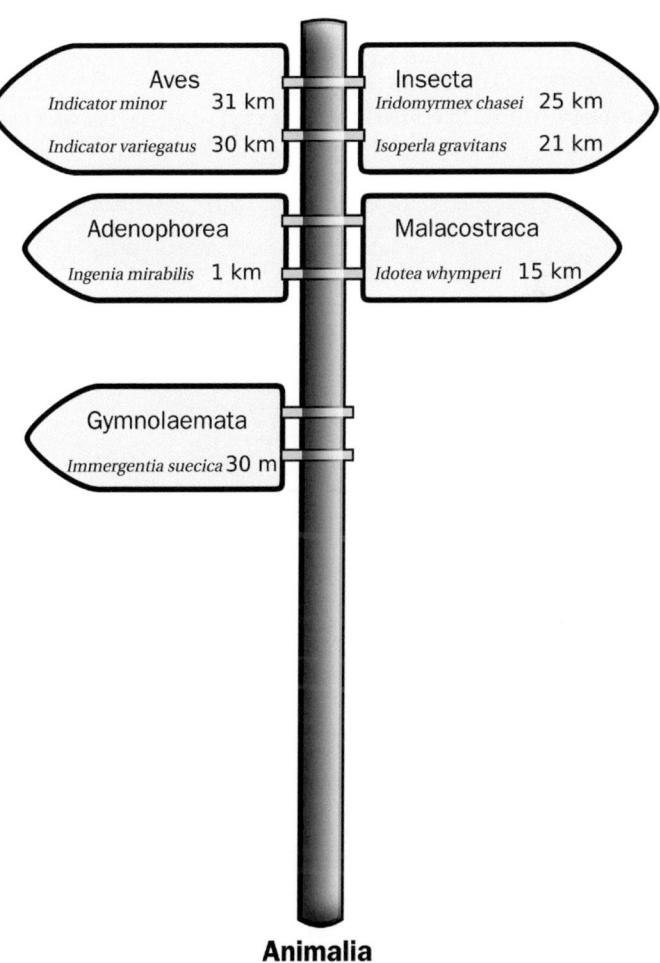

Animalia

Idotea whymperi

generally the wrong answer to udotea javensis (q.v.). Real scientists drink coffee.

Immergentia suecica

the neurotic inbred lab animal that drops dead when the fire alarm goes off.

Indicator minor minor

the very small hint supporting the pet theory of your boss. Many years and theses are spent chasing Indicator minor minor until he has finally retired.

Indicator variegatus

the results indicating that the pet theory of the supervisor is wrong. In contrast to minor minor, several indicator variegatus will severely reduce the chance of getting your contract renewed.

Ingenia mirabilis

the solution to a long-standing problem that pops out of thin air while under the shower. Ingenia quite often spends a considerable time in the disguise of a dysidea amblia (q.v.).

Iridomyrmex chasei yaglooensis

the messy catch of your escaped lab animal.

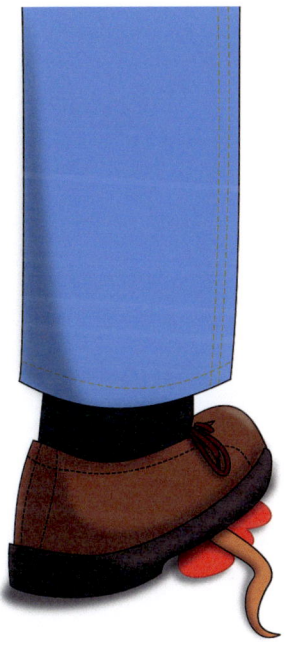

Isoperla gravitans

the non-magnetic beads in the batch of magnetic ones.

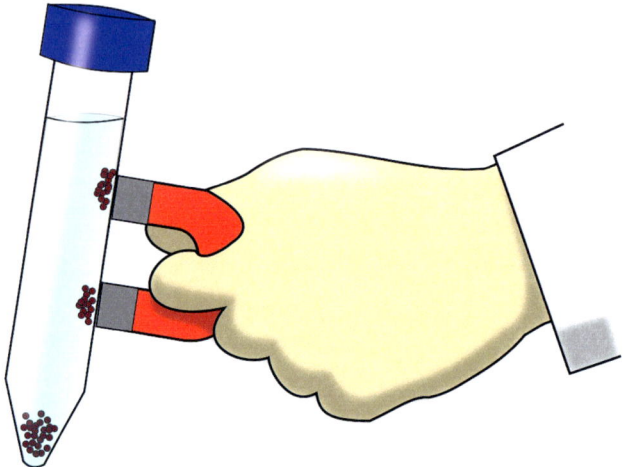

J

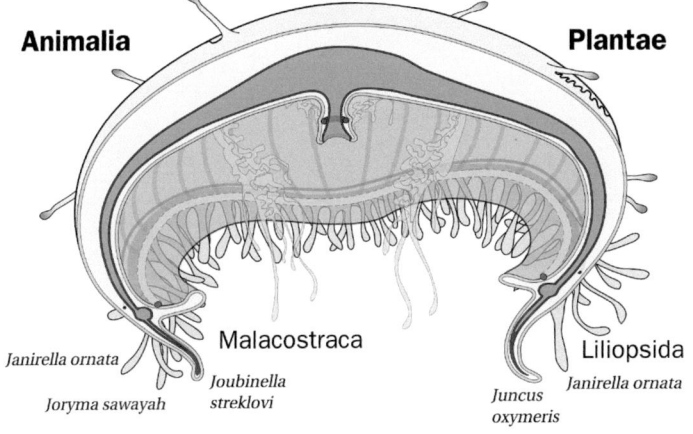

Animalia

Plantae

Malacostraca

Liliopsida

Janirella ornata

Janirella ornata

Joryma sawayah

Joubinella streklovi

Juncus oxymeris

Janirella ornata

the equivalent of crop circles in the lab. Strange streaks and circular markings appearing on the floor over night. These are often encrypted attempts of the cleaning personnel to communicate that hazardous waste material should be disposed of properly.

Joryma sawayah

the hasty remark you make to your colleagues before you have to leave for the congress. Somehow it's difficult keeping track of time when still working on the presentation.

Joubinella streklovi

the sci-fi fan that initiates all ageing research experiments with parting his hands and the phrase "live long and prosper".

Juncus oxymeris

the problem with explaining the existence of non-coding genes to the general public.

Juncus patens

the patent submitted to polish up one's CV.

K

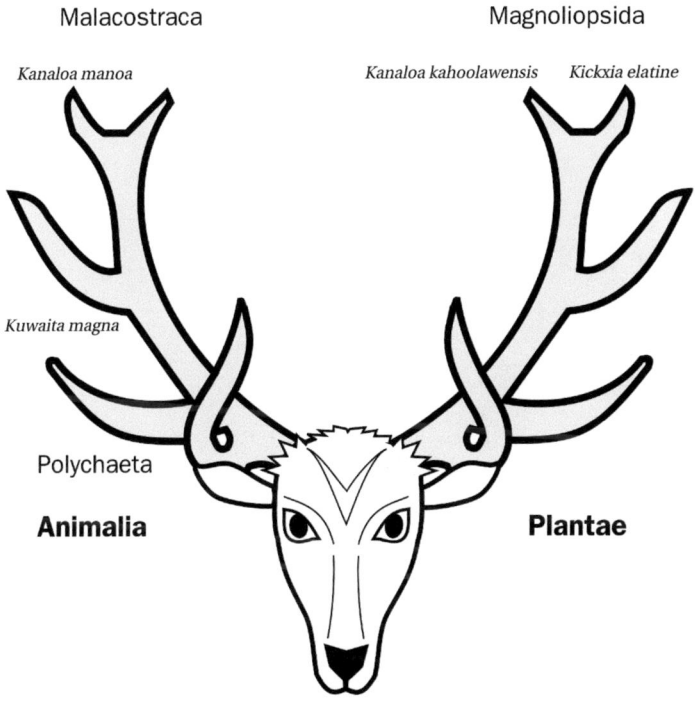

Malacostraca

Magnoliopsida

Kanaloa manoa

Kanaloa kahoolawensis   *Kickxia elatine*

Kuwaita magna

Polychaeta

Animalia

Plantae

Kanaloa kahoolawensis

scientific congresses taking place on tropic islands. For some inexplicable reason those are also best attended by the luminaries in your field of research.

Kanaloa manoa

person in charge of travel expense funds who gets to decide who may travel to a kahoolawensis (q.v.).

Kickxia elatine

the extra buzz you get by proving your supervisor wrong with a dashing experiment.

Kuwaita magna

the experiment that gets delayed because you're still waiting for the shipment despite repeated reassurance from the supplier that it should arrive anytime soon.

L

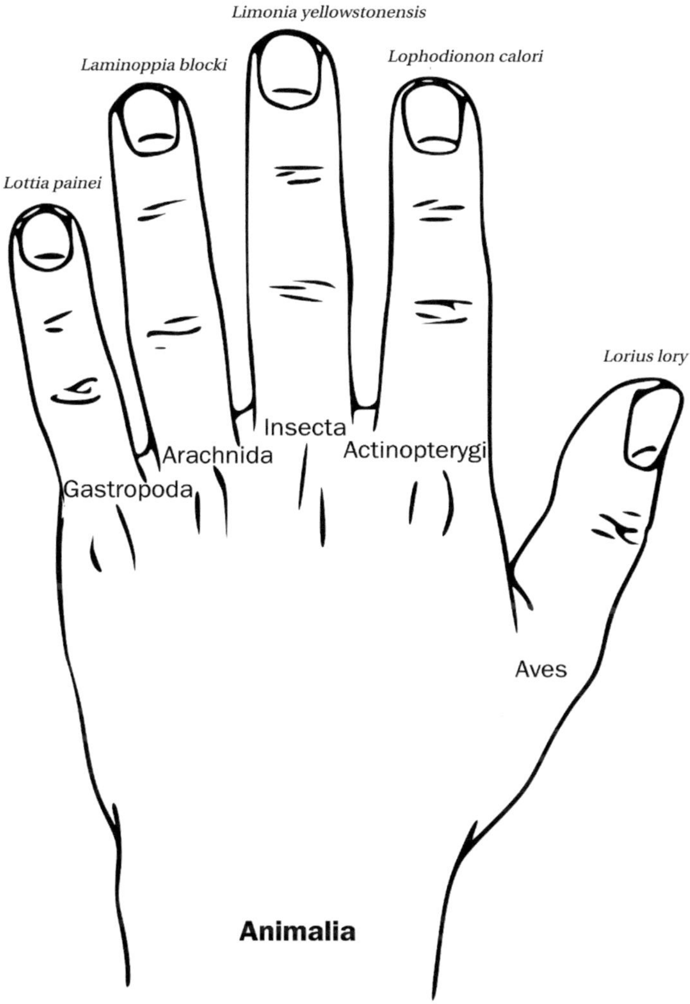

Lottia painei

Laminoppia blocki

Limonia yellowstonensis

Lophodionon calori

Lorius lory

Insecta

Arachnida

Actinopterygi

Gastropoda

Aves

Animalia

Laminoppia blocki
the magical solution for every experiment. It contains an undisclosed and highly variable amount of components like BSA, polyIdC, Tween and other substances with interesting acronyms. Laminoppia is sold by companies with promising names and incredible properties to sell their snake oil with staggering profits.

Limonia yellowstonensis
the awkward colour change of your buffer after it came back from the autoclave.

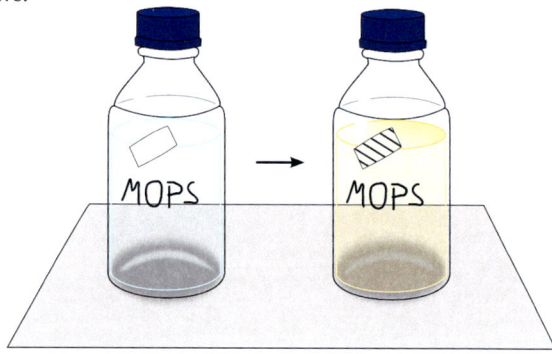

Lophodionon calori
the amount of calories burned while waiting for the ultracentrifuge to spin down without brakes.

Lorius lory jobensis
the [g]noble reason why scientists will work for low salary on limited contracts.

Lottia painei
the state you go through when stumbling upon a recent paper describing exactly the results you're still struggling to obtain.

M

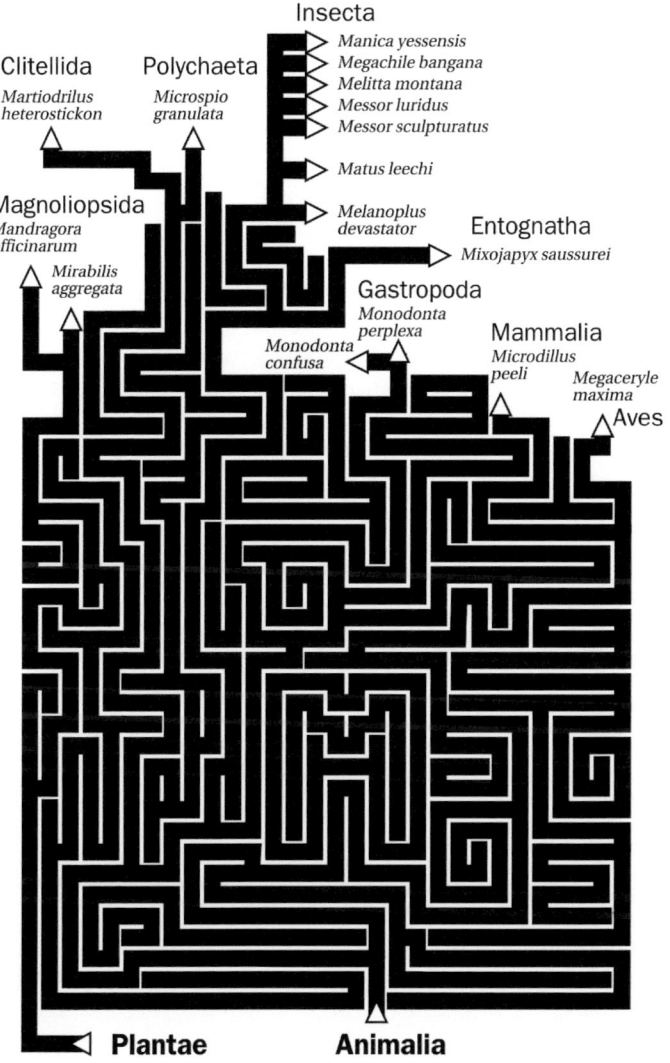

Insecta
- *Manica yessensis*
- *Megachile bangana*
- *Melitta montana*
- *Messor luridus*
- *Messor sculpturatus*
- *Matus leechi*
- *Melanoplus devastator*

Clitellida
Martiodrilus heterostickon

Polychaeta
Microspio granulata

Entognatha
Mixojapyx saussurei

Magnoliopsida
Mandragora officinarum

Mirabilis aggregata

Gastropoda
Monodonta perplexa

Monodonta confusa

Mammalia
Microdillus peeli

Megaceryle maxima

Aves

Plantae   Animalia

Mandragora officinarum
having to do the pesky paperwork and dealing with regulations instead of lab work. The longer you're into the science business the greater the mandragora.

Mangelia acloneta acloneta
the prehistoric bacteriophage lab contamination that continues to claim your coli strains to this day.

Manica yessensis
the affirmative behaviour often observed in foreign students with poor language skills. Any obtained information should be confirmed by reconstructing the previous question in order to yield a negative answer.

Marginella virginiana
the profit made if you sell your messor luridus virginalis (q.v.) via the internet.

Martiodrilus heterostickon
the recycled sample box with contradicting labels on it.

Matus leechi
the colleague that asks once whether he can borrow something quickly. Then continues to take it away repeatedly without any further notice.

Megaceryle maxima gigantea
the huge comb in the back of a drawer that makes you speculate that electrophoresis chambers exceeding the size of a metre must have once actually been used in the lab. Such a proposed megascience may have existed shortly after biochemistry split from the classical chemistry branch.

Megachile bangana

the inexplicable build up of tension when pressing the electroporator buttons.

Melanoplus devastator conspicuus

the slight facial flush of the colleague that nonchalantly tells you "it wasn't me" when you ask for the one who knocked over the rack of your open samples.

Melitta montana

the pile of name badges from past meetings starting to take over the remaining space of your desk.

Messor luridus virginalis

the artefact on a western blot, microarray or other visual experimental result that even slightly resembles the Virgin Mary.

Messor sculpturatus
the tidiness-challenged colleague that accumulates impressive pieces of modern art on his bench.

Microdillus peeli
the thin section obtained by a microtome that came off either too small or crinkled to be handled properly.

Microspio granulata
the picture obtained at the very highest resolution that can be achieved by a given microscope. Other than in some TV series there are certain physical laws that prove very difficult to be bent. One further nudge at the scale may lead to a chrysotrichia (q.v.) and loss of precious sample.

Mirabilis aggregata
the typical but unreproducible behaviour of your protein of interest.

Mixojapyx saussurei
when you're convinced that you did pipet the enzyme into the reaction mixture only to find out later that, actually, you didn't.

Monodonta confusa
one colony more comparable with the negative control after the transformation.

Monodonta perplexa
one colony less comparable with the negative control after the transformation.

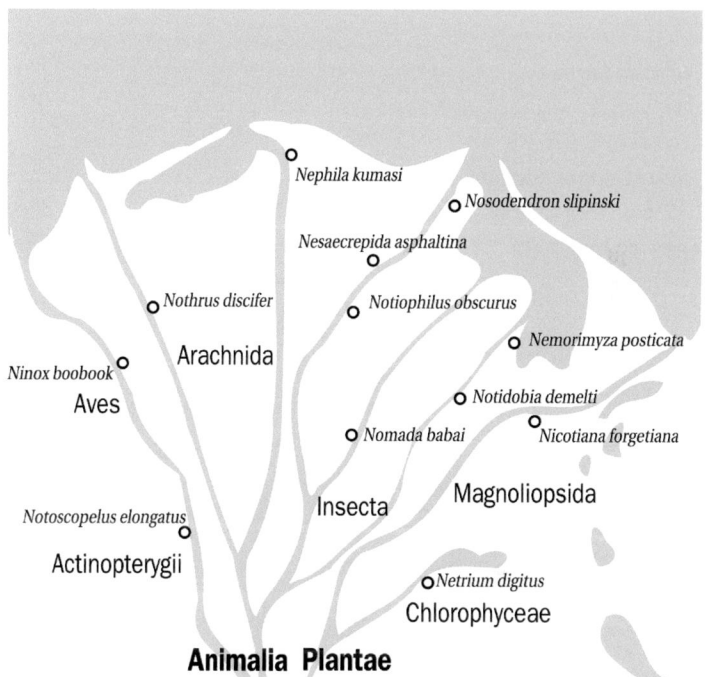

Nephila kumasi

Nosodendron slipinski

Nesaecrepida asphaltina

Nothrus discifer

Notiophilus obscurus

Arachnida

Nemorimyza posticata

Ninox boobook

Aves

Notidobia demelti

Nomada babai

Nicotiana forgetiana

Insecta

Magnoliopsida

Notoscopelus elongatus

Actinopterygii

Netrium digitus

Chlorophyceae

Animalia Plantae

Nemorimyza posticata

the chaos that ensues when you forget to remind yourself to obtain more post-its.

Nephila kumasi

the stains found on a gel, lab coat, fingers, and forehead.

Nesaecrepida asphaltina

the 100x stock solution of coffee found in the percolator that was freshly prepared over night. Excess stock solution is collected and used for the annual filling of potholes in the drive to the institute.

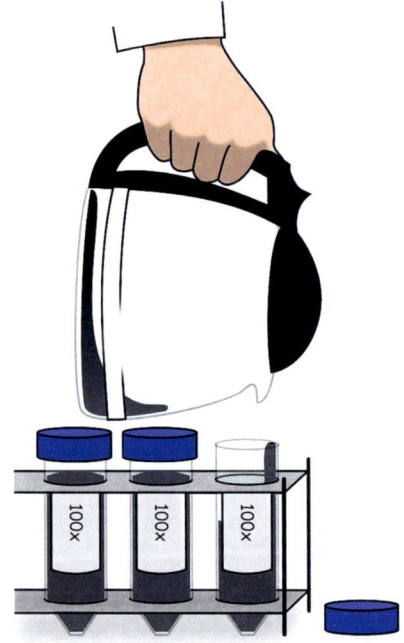

Netrium digitus constrictum

the problem encountered each time raw data files need to be transferred from a remote computer while the virus scanner is still active.

Nicotiana forgetiana

the 'smoking gun' found in the lavatory. Quite often the reason for an asio flammeus domingensis (q.v.).

Ninox boobook

yet another obsolete dictionary for molecular biology. Ninox is a very likely birthday present for a dear colleague after other options like pocket calendars, gift vouchers, and text markers have been exhausted over the years. It is also quite useful as a wedge for the short leg of a wobbly table.

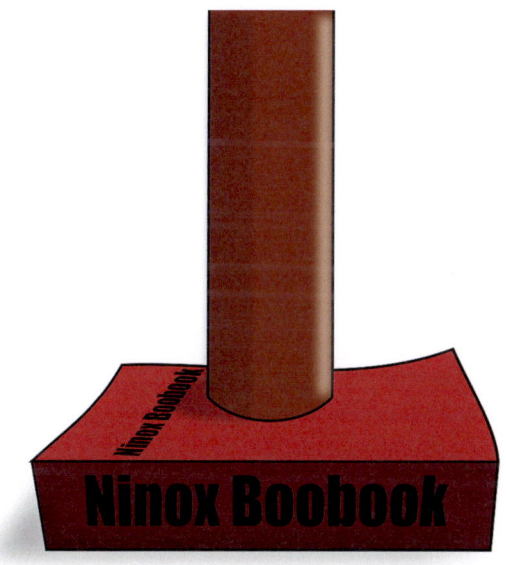

Nomada babai

the final result late in the evening that the entire long term experiment was wasted by an oopsacas minuta (q.v.). The right thing is to call it a day and go home.

Nosodendron slipinskii

the puddle forming under a leaky safety shower.

Nosodendron slipinskii obtectum

the puddle under the safety shower hidden by soaked tissue.

Nothrus discifer

the futile attempt to make sense of the experiment you entered in the lab book just a month ago.

Notidobia demelti

when global warming had to start on your shipment somewhere on its way so that all samples can finally be pronounced dead on arrival.

Notiophilus obscurus

the guy that likes to adhere to the least likely explanation for a given phenomenon. Notiophilus is quite often consulted as a supposedly reliable scientific reference for politicians, religious groups, or flat-earthers.

Notoscopelus elongatus elongatus

the guy that pretends to ask a short question at the end of a lecture but continues to ramble for another 10 minutes.

O

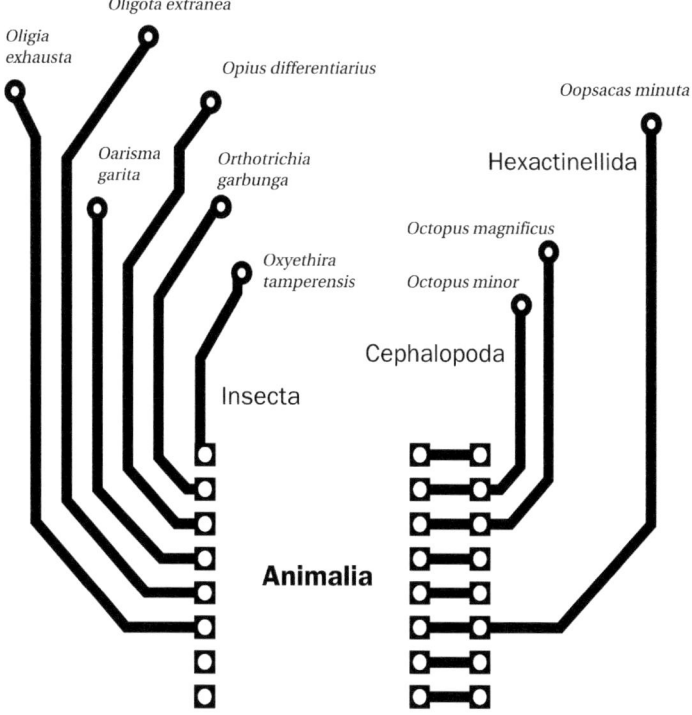

Oligota extranea

Oligia exhausta

Opius differentiarius

Oarisma garita

Orthotrichia garbunga

Oxyethira tamperensis

Insecta

Animalia

Oopsacas minuta

Hexactinellida

Octopus magnificus

Octopus minor

Cephalopoda

Oarisma garita

the illness contracted when deadlines are due. Symptoms can range from mild amnesia of where relevant documents are found up to previously unheard of life threatening conditions as stated on the medical certificate.

Octopus magnificus

the tranquillity in the lab at 8 in the morning.

Octopus minor typicus

the happiness of leaving the lab before 8 in the evening.

Oligia exhausta

the very end of cling film, autoclave tape, parafilm, and ...PhD thesis.

Oligota extranea

the bands in an electrophoresis gel not belonging to your desired product.

Oopsacas minuta

the very moment you realise that you really shouldn't have pipetted this into that.

Opius differentiarius

the rare feeling that you actually achieved something. Opius should be savoured thoroughly along with bubbly drinks to last through the rough patches.

Orthotrichia garbunga

the mess created when you tip over the waste bin on your cleanbench.

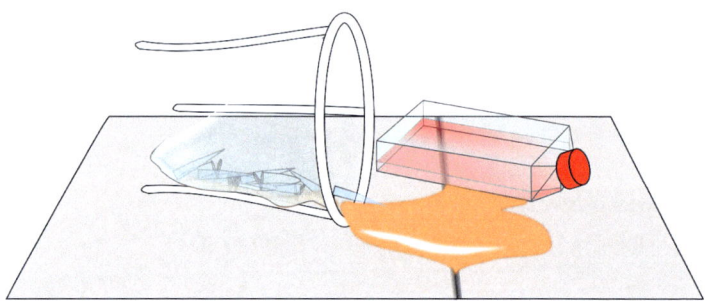

Oxyethira tamperensis

results obtained under high pressure and supervisor's expectations in the final stages of a PhD thesis. Oxyethira also leads to loss of raw data and selective amnesia as soon as the PhD graduate has left the lab.

P

Creation soup

Ingredients P

contains

<u>Animalia</u>

Insecta

Coleoptera *(Petalium alaseriatum, Pericompus elegantulus, Pachybrachis chaoticus, Ptinus fur)*

Diptera *(Pipiza quadrimaculata, Pelmatops fukiensis)*

Hemiptera *(Parthenicus deleticus, Parthenicus obsoletus)*

Hymenoptera *(Pheidole ridicula, Pseudomyrmex pictus castus)*

Leptidoptera *(Pseudoplusia includens)*

Phthiraptera *(Platygaster flabellata)*

Siphonaptera *(Pulex irritans)*

Thysanoptera *(Pseudothrips inequalis)*

Malacostraca

Amphipoda *(Parathemisto oblivia)*

Isopoda *(Porcellio transmutatus)*

Actinopterygii

Scorpaeniformes *(Paraliparis csiroi)*

Atheriniformes *(Pseudomugil inconspicuus)*

Amphibia

Caudata *(Pseudoeurycea mixcoatl)*

Anura *(Psychrophrynella usurpator)*

Bivalvia

Pholadomyoida *(Pandora granulata, Pandora inflata, Periploma papyratium)*

Gastopoda

Archaeogastropoda *(Puncturella agger)*

Heterostropha *(Pseudomalaxis centrifuga)*

<u>Plantae</u>

Liliopsida

Orchidales *(Pelexia adnata)*

Magnoliopsida

Scropholariales *(Pinguicula vulgaris)*

Pachybrachis chaoticus

the colleague that labels all his samples with numbers and keeps them in an open box in the freezer. The problem only starts to surface when any sample is suddenly needed by the end of the thesis.

Pandora granulata

the vast space of your shipment occupied by a strange fluffy material. Some physicist see this as the form dark matter takes on earth.

Pandora inflata

the shipments that follow the iterative Matyroshka principle yielding the item of interest in the very last tiny box.

Paraliparis csiroi

the colourful steaming liquids and equipment eye candy shown in a lab of an investigative tv series that look exactly unlike anything you've ever come across in real life.

Parathemisto oblivia

the elusive fraction where most of your sample vanishes during chromatography.

Parthenicus deleticus
the component of any kit that is used up first and not available separately.

Parthenicus obsoletus
the surplus of any kit that accumulates over time and crams the cupboards.

Pelexia adnata
the difficulty of reading and writing only four different letters. A severe impediment to overcome for any prospective molecular biologist.

Pelmatops fukiensis
the decisive sweep to move the heaps of scrap paper, communication, and long sought manuscripts into the bin before going on vacation.

Pericompus elegantulus
the most effective evasive manoeuvre in a lab meeting. Bringing up minor IT issues can completely derail the discussion from less comfortable topics like deadlines or lab duties.

Periploma papyratium
all failed experiments that don't make it into the final thesis. Periploma usually exceeds the final version by orders of magnitude, resulting in environment-friendly hardcopies.

Petalium alaseriatum
the exact time point during a presentation that laser pointer batteries are designed to die on.

Pheidole ridicula
the last droplet coming from the gravity flow column that simply refuses to drop.

Pinguicula vulgaris
the lab instruments emitting a high pitched tone. The tone can convey various meanings like the beginning, commencement, or end of an experiment, the need for human interaction, or general malfunction. A fortune could be made if downloadable tones were made available to tell them apart.

Pipiza quadrimaculata
the carefully balanced minimal medium some of your colleagues seem to live on every lunchtime.

Platygaster flabellata
the surprise question posed by non-scientists late at a party: „so what kind of research do you do, actually?" One good answer may be „Nobody expects the Spanish inquisition."

Porcellio transmutatus
the petrified plasticware sometimes found in less frequented storage compartments in the lab. A still unknown, probably light-sensitive mechanism slowly turns petri dishes, syringes, pipettes and small receptacles into glassware. Extant date markings on the retrieved material indicates that the process may have taken more than 20 years.

Pseudoeurycea mixcoatl
the sudden awareness of wearing the wrong lab coat during the annual security check or other official occasion.

Pseudomalaxis centrifuga
the accidental asymmetric sample placement in the centrifuge rotor resulting in a surnia (q.v.).

Pseudomugil inconspicuus

the magic cloaking device that allows you to blend in with your fellow scientists during congresses in order to avoid situations like galactia smalli (q.v.).

Pseudomyrmex pictus castus

the credibility of a press release depicting the institute's director sporting a lab coat and protective goggles gazing intensely at some colourful liquid in a reaction tube.

Pseudoplusia includens

the samples one day beyond expiration date. Mostly unproblematic, but now your experiments leave terra firma and enter the realm of lower reproducibility. Depending on your budget and/or time frame, pseudoplusia is a necessary evil.

Pseudothrips inequalis

the sample obtained from a partner in a large research consortium that turns out to be exactly unlike what you were told.

Psychrophrynella usurpator

the few unquestioned potentates left in the transition from the feudal system to the generally powerless individuals in life sciences. The last strongholds existing almost exclusively in the realms of medical sciences.

Ptinus fur

the strange shape powdery samples take when you're wearing insulating shoes and gloves.

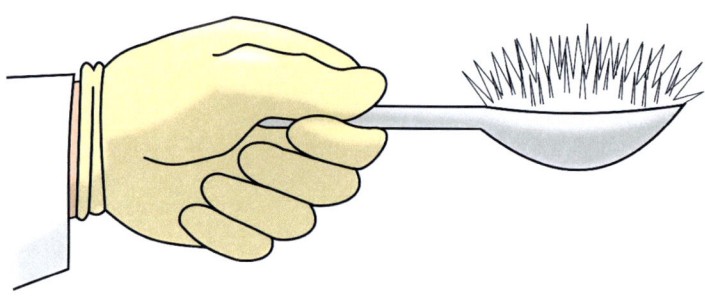

Pulex irritans

an itch you can't scratch right during an important experiment. Pulex usually affects your nose while handling radioactive isotopes.

Puncturella agger

the inevitable leak that causes your gel to escape the casting before it can polymerise.

Q

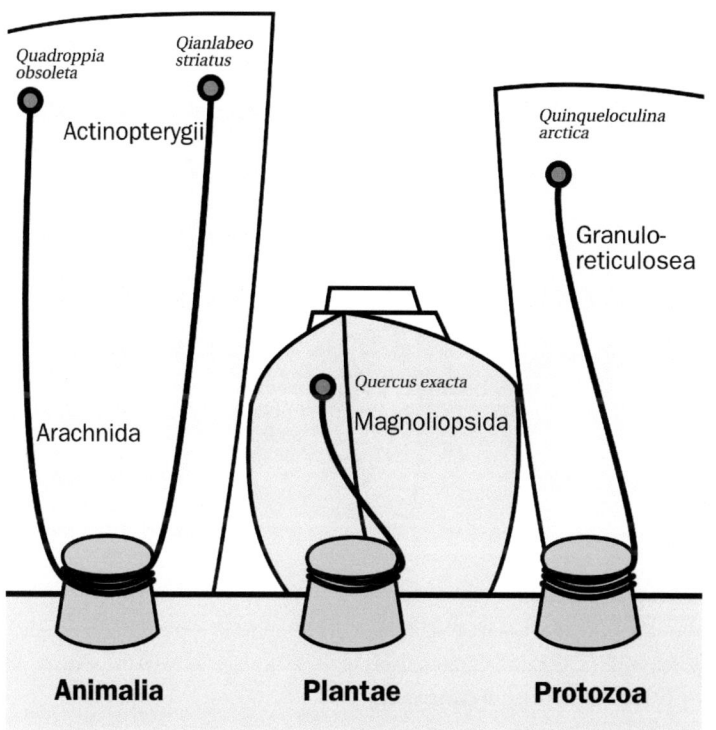

Qianlabeo striatus

the problem with the generation Q of researchers. If no kit for a given experiment is available it becomes a mission impossible.

Quadroppia obsoleta

the adhesive tape for labelling of important samples that is especially designed to come off in the freezer.

Quercus X exacta

the colleague that always calculates and measures to the umptieth decimal place while at the same time reliably exceeding experimental error margins by at least two orders of magnitude.

Quinqueloculina arctica

the one glycerol stock culture of that refuses to grow despite several attempts to revive it.

R

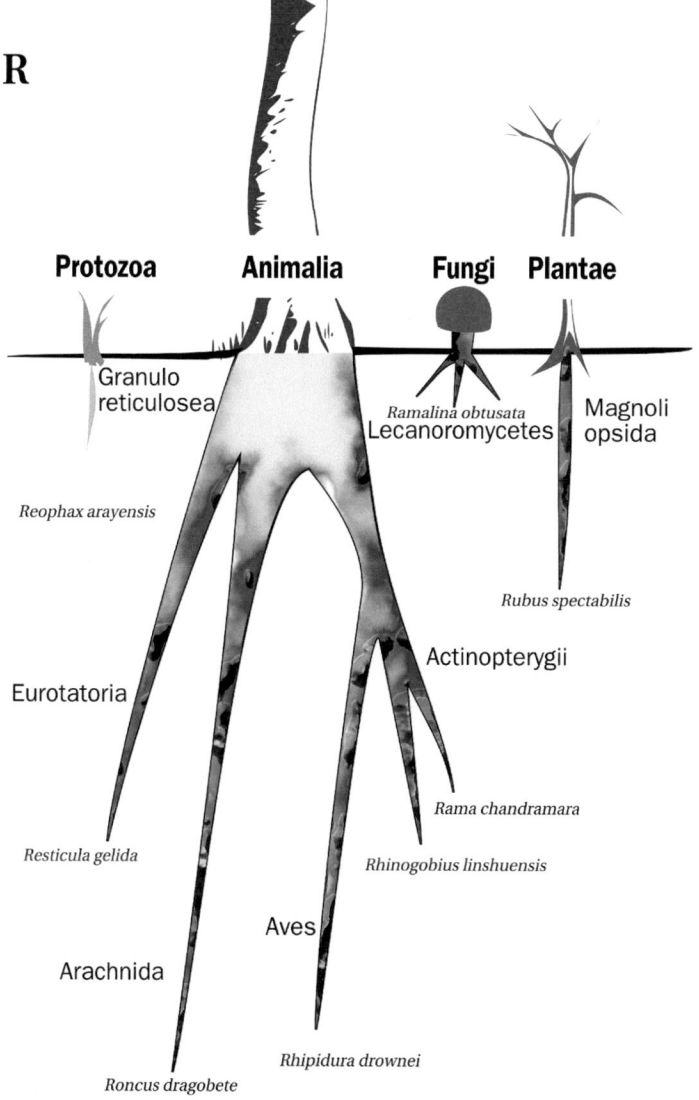

Protozoa   **Animalia**   **Fungi**   **Plantae**

Granulo
reticulosea

Ramalina obtusata
Lecanoromycetes

Magnoli
opsida

Reophax arayensis

Rubus spectabilis

Actinopterygii

Eurotatoria

Rama chandramara

Rhinogobius linshuensis

Resticula gelida

Aves

Arachnida

Rhipidura drownei

Roncus dragobete

Rama chandramara

the Asian deity of high-throughput that can hold and handle eight pipettes simultaneously (see illustration of ancient tapestry). Newer depictions display 12er multichannel pipettes in every hand, making it more microtitre plate compatible.

Ramalina obtusata

the modern and more stylised statue of Rama chandramara found in almost every lab around the world. Ignorant western scientist only know and use it as a pipette holder.

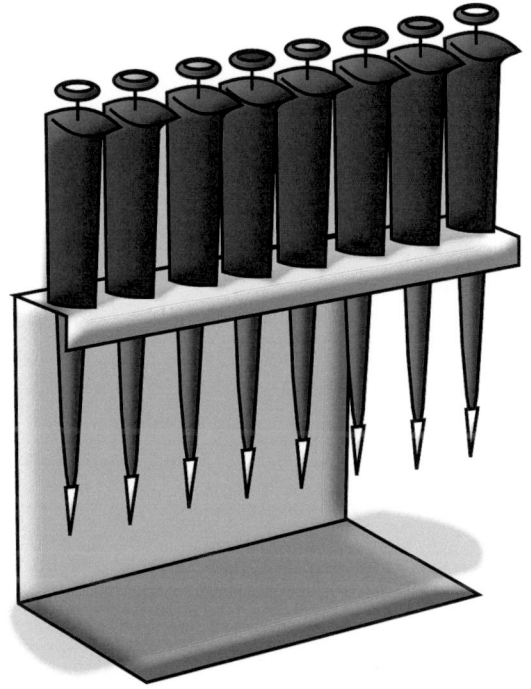

Reophax arayensis

the huge blotch or streak covering most spots in a grid including the duplicates. Reophax is found exclusively on slides containing the most interesting data.

Resticula gelida

the very last unit of precious enzyme spun down from the walls of the tube.

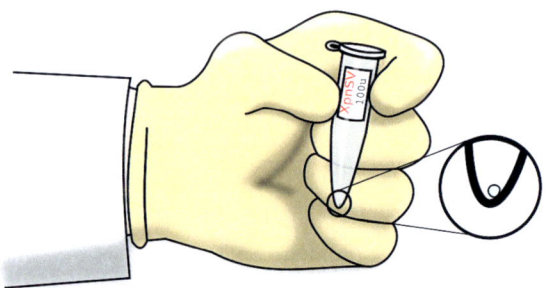

Rhinogobius linshuiensis

the eastern art to perfectly arrange the stack of membranes in the semi-dry western blot to allow an even flow of electric currents.

Rhipidura drownei

the dislodged pipette tip found in liquid cultures. Often dispensed deliberately as to keep track which samples already got inoculated.

Roncus dragobete

the guy that always talks about an important experiment whenever you meet him but never gets around to actually doing it.

Rubus spectabilis

the moment you realise that you burned your face with the transilluminator wearing the wrong protective goggles.

S

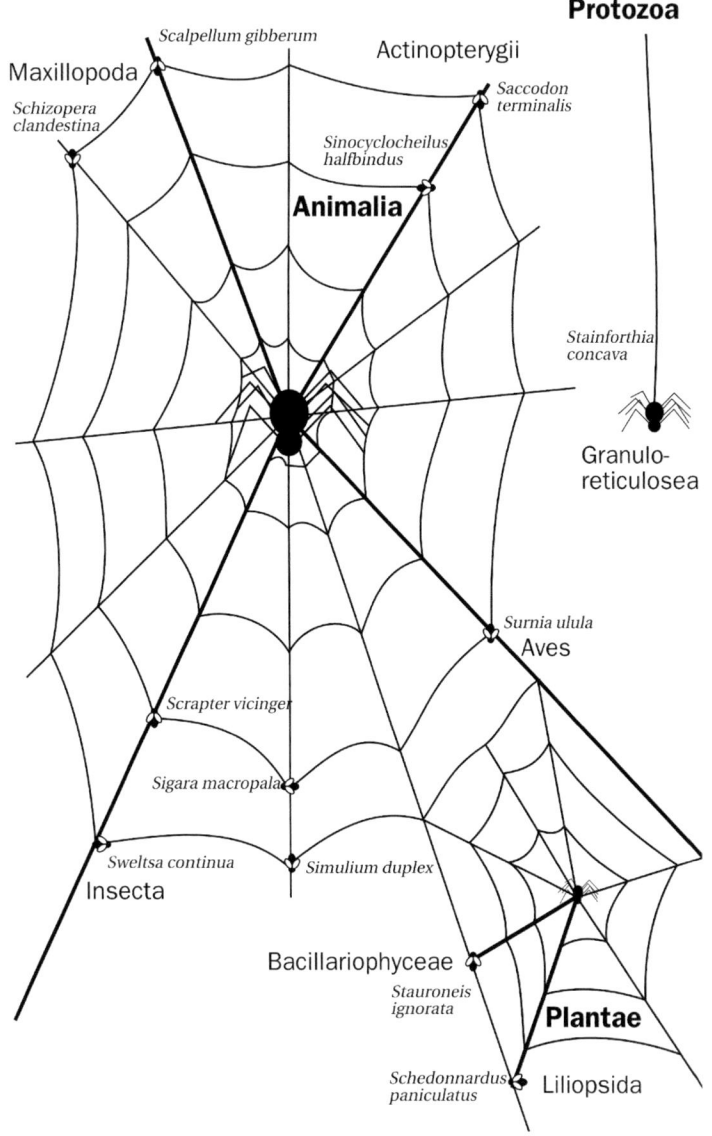

Saccodon terminalis

the single sequencing ambiguity in the cloning primer area of your expression construct later to turn out to constitute a stop codon.

Scalpellum gibberum

the sticky stuff accumulating on a frequently used blade.

Schedonnardus paniculatus

the experiment you perform out of schedule while desperately waiting for your ordered samples to arrive.

Schizopera clandestina clandestina

the top secret experiment you won't even tell your other paranoid self about.

Scrapter vicinger

the special tool that is designed to pick up gels from electrophoresis chambers and turn them into challenging jigsaw puzzles.

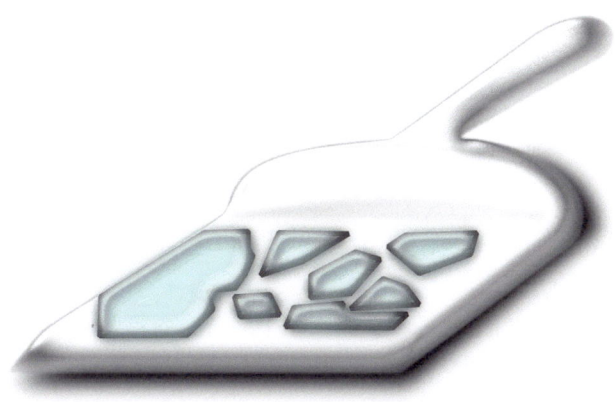

Sigara macropala

the makeshift ashtray found at any of the institute's emergency exits overflowing with discarded cigarette butts.

Simulium duplex

the control that you deem unnecessary, but later turns out to be indispensable for proper evaluation.

Sinocyclocheilus halfibindus

the odd antibody that will only work in every second experiment. However, parallel duplicates always turn out the same.

Stainforthia concava

the crescent-shaped coffee stains left behind on all important documents you're working on. It serves as an unofficial verification stamp for your superior that you worked thoroughly on papers, manuscripts, lab book, and sudoku sheets.

Stauroneis ignorata

the author in a publication that didn't co-write, contribute, finance, or endorse, and most likely wouldn't remember any scientific details nor title word if asked.

Stenus vacuus

the extra force needed to reopen the -80 just after you closed it.

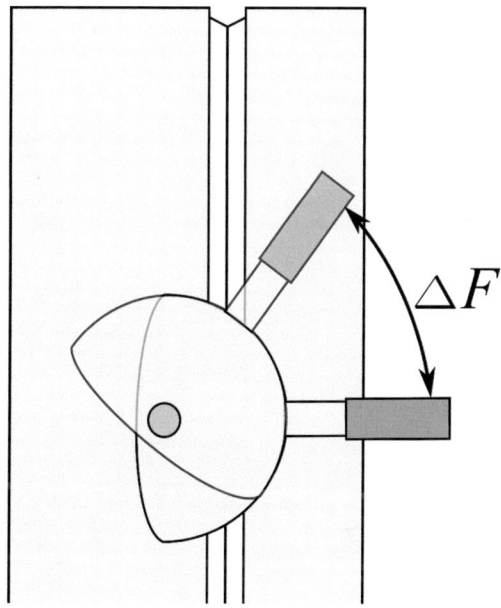

Surnia ulula ulula

the wailing sound that tells you that the centrifuge is definitely out of balance.

Sweltsa continua

the miraculous breeze around the precision scales that persists even when all doors are closed.

T

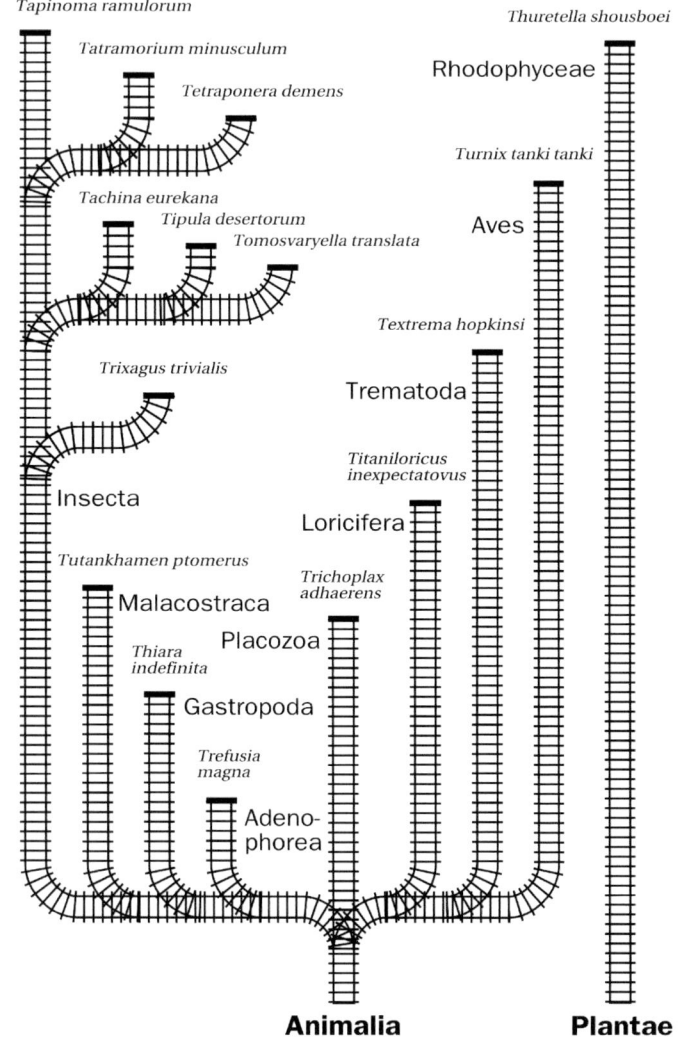

Tapinoma ramulorum

Tatramorium minusculum

Tetraponera demens

Rhodophyceae

Thuretella shousboei

Tachina eurekana

Tipula desertorum

Tomosvaryella translata

Turnix tanki tanki

Aves

Trixagus trivialis

Textrema hopkinsi

Trematoda

Insecta

Titaniloricus inexpectatovus

Loricifera

Tutankhamen ptomerus

Malacostraca

Trichoplax adhaerens

Placozoa

Thiara indefinita

Gastropoda

Trefusia magna

Adeno-
phorea

Animalia

Plantae

Tachina eurekana

the joy when you finally discover your long coveted clone.

Tapinoma ramulorum inrectum

the experiment that you managed to trash thoroughly, short of setting the entire lab on fire. Tapinoma is one of the few experiments that are best never repeated unless you want to quit research for good.

Tetramorium minusculum

the amount of product retained after more than four sequential synthesis and purification steps.

Tetraponera demens

the fourth time in a row you forgot where you put your sample. Time to get a good night's sleep.

Textrema hopkinsi

the widespread belief that quantity can cover up poor results. A frequent reason that lengthy manuscripts are turned down after some sleep-inducing chapters were read by referees.

Thiara indefinita

the typical makeshift theory. If something doesn't seem to fit right you'll turn it around and postulate the opposite.

Thuretella schousboei

the debilitating compulsive disorder that afflicts some scientists occasionally when things finally turn out as expected. Thuretella leads to rapid uttering of vocabulary like eureka, bingo, wow, yesss, cool etc. and may last from minutes up to an entire day.

Tipula desertorum

the pipette tip that always becomes unstuck from your multichannel.

Titaniloricus inexpectatovus

the serendipitous discovery that changes everything including getting free invitations to a kahoolawensis (q.v.). One day titaniloricus will happen to you, too.

Tomosvaryella translata

all valuable information that gets either lost or distorted in the communication with your foreign research partner.

Trefusia magna

the experimental protocol that is always dead certain to work, but simply refuses to work in your hands even for the third time in a row. Trefusia may be caused by a Tutankhamen (q.v.).

Trichoplax adhaerens

the fuzzy often colourful horizontal line on the inner wall of an old tabletop centrifuge. Usually the 'hottest' item in the isotope lab.

Trixagus trivialis

the projected duration of a perfectly simple experiment plus another day, multiplied by two, and an additional PhD thesis.

Turnix tanki tanki

any part in the shaking incubator that is prone to come off and jam the mechanism in a place which is most difficult to reach.

Tutankhamen pteromerus

the curse contracted when opening an ancient viscosia (q.v.). It causes several years of experimental misfortune. The main reason why those samples are better left untouched so future generations can deal with them properly.

U

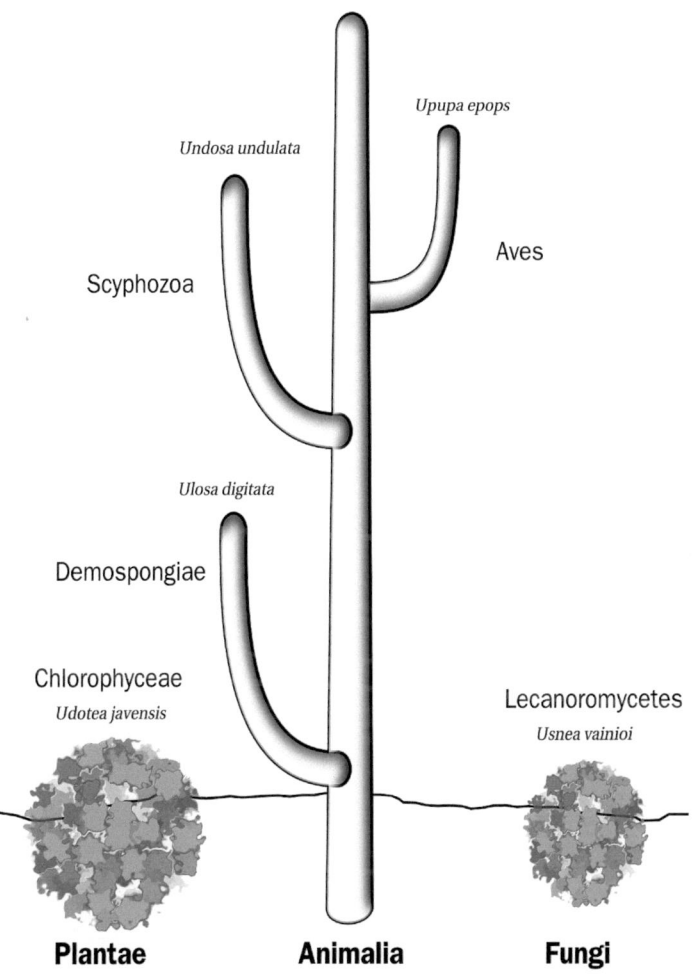

Upupa epops

Aves

Undosa undulata

Scyphozoa

Ulosa digitata

Demospongiae

Chlorophyceae

Udotea javensis

Lecanoromycetes

Usnea vainioi

Plantae   **Animalia**   **Fungi**

Udotea javensis

an important question in life science. Are you a tea or coffee person? As many colleagues are wary of tea drinkers, you answer with javensis whenever asked at meetings.

Ulosa digitata

the headcrash claiming all your primary data and manuscripts on your hard drive. It normally happens just when you are thinking more strongly about how important a backup may be.

Undosa undulata

the immediate attempt to rescue the sample directly after the oopsacas minuta (q.v.) occurred.

Upupa epops epops

anything that doesn't belong into the microwave in hindsight.

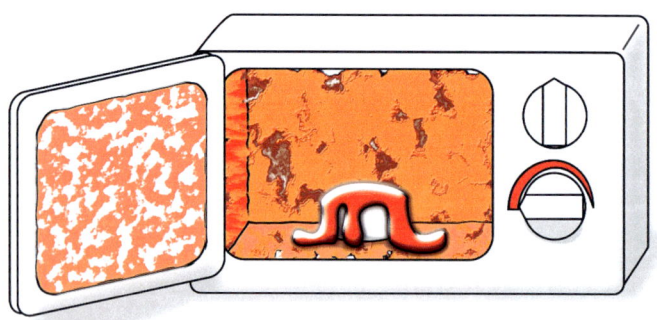

Usnea vainioi

the device bought at the last asellus quicki (q.v) that hasn't been touched since.

V

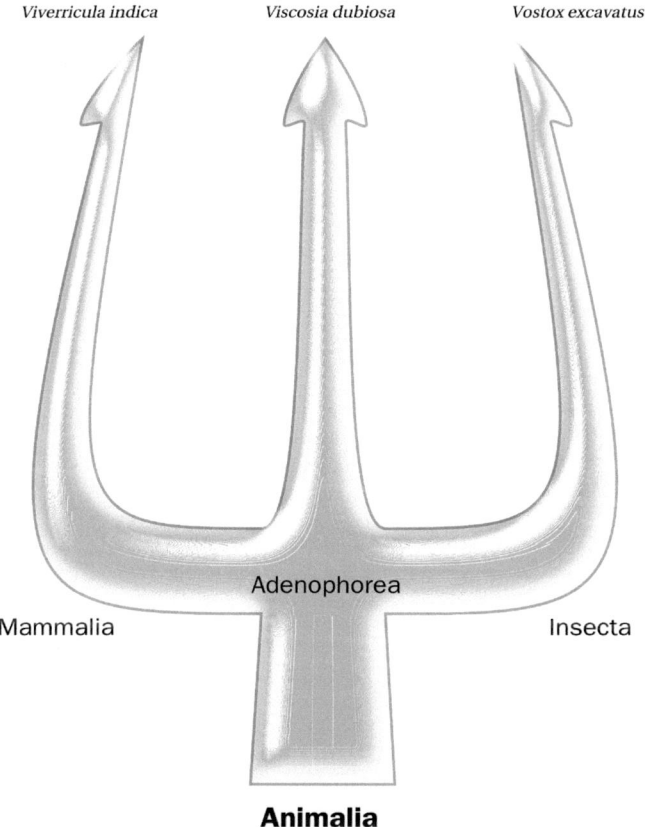

Viscosia dubiosa

the cryptically labelled possibly important sample in the fridge passed on for generations.

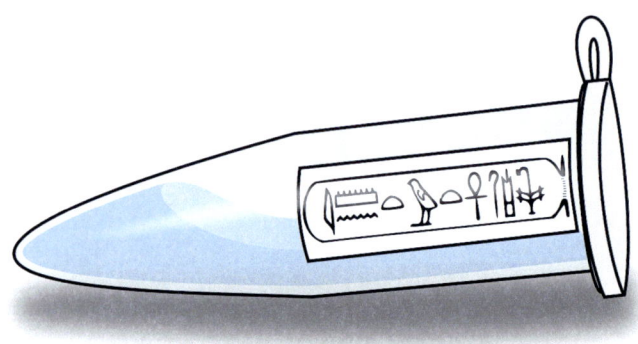

Viverricula indica deserti

best indication for imminent start of the weekend given by regular staff on a Friday afternoon.

Vostox excavatus

the frantic search for your sample in the overflowing -80 freezer.

W

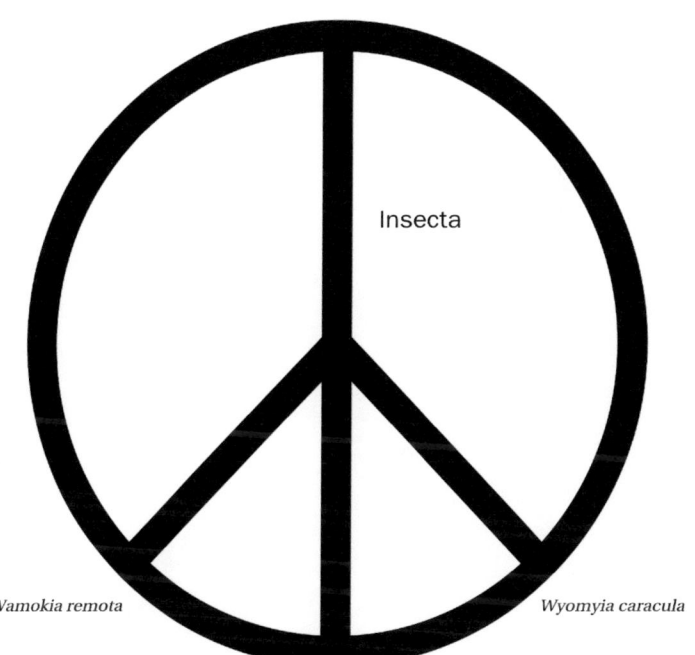

Wamokia remota

the distressing wake up call by mobile phone in a dead silence of a boring plenary lecture. Caring colleagues will always offer a refreshing wamokia when they see you nodding off.

Wockia asperipunctella

the unexpected bizarre phenotype obtained from your mutagenesis screen. Wockia is also a rare occasion to freely express your creativity by coining an even more ridiculous name for the observed mutation.

Wyeomyia caracula

expression of awe uttered by innocent bystanders witnessing a tapinoma (q.v.) unfold.

Y

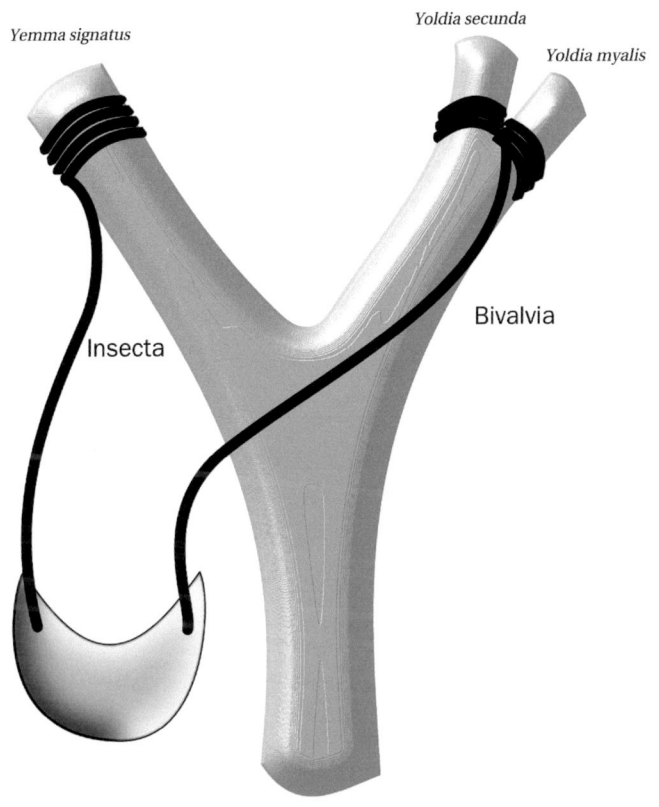

Yemma signatus

Yoldia secunda

Yoldia myalis

Insecta

Bivalvia

Animalia

Yemma signatus

the cramp that increasingly affects your signature when signing 10 receipt forms in a row.

Yoldia myalis

the offset to your yoldia secunda (q.v.) achieved in real life that determines the reproducibility of the given experiment. If myalis exceeds the secunda by a certain value, experienced scientists simply cease to care as long as a qualitative result is sufficient.

Yoldia secunda

the perfect duration for pipetting the enzyme, mixing the sample, closing the lid, and putting it into the incubator for all the samples you're supposed to be handling simultaneously.

Yucca constricta

the stench escaping from a disused sink with a dried out syphon that connects directly to the drain of the chemistry department.

Z

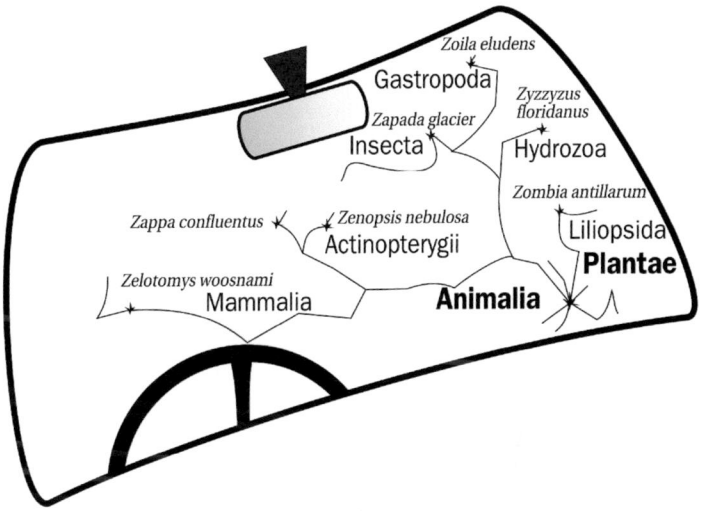

Zoila eludens
Gastropoda
Zyzzyzus floridanus
Zapada glacier
Insecta
Hydrozoa
Zombia antillarum
Zappa confluentus
Zenopsis nebulosa
Liliopsida
Actinopterygii
Plantae
Zelotomys woosnami
Mammalia
Animalia

Zapada glacier

the final status of samples stored in the freezer. Samples are archived chronologically and tightly held together without any tedious boxes or racks by a slow but gradual glaciation process guided by a built in feature found in all freezers. Zapada glacier is designed to serve as a time capsule so future generations of researchers can retrieve unperturbed samples from drill cores.

Zappa confluentus

the little extra salt in your electroporation sample. Frequent zappa is also the reason for the development of a megachile (q.v.) condition.

Zelotomys woosnami

all authors appearing neither at first nor last position on a publication.

Zenopsis nebulosa

the increased vapour and water condensation in the cold room indicating that it may already be summer outside.

Zoila eludens

the nondescript gadget left over from your predecessor no one can guess what its use may be.

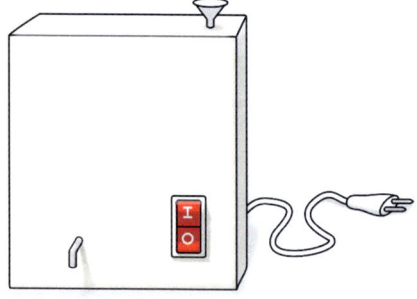

Zombia antillarum
the kind of cloning vector that bounces back without insert despite triple digest, dephosphorylation, gel excision, toxic cassette, cursing, and potent voodoo rituals.

Zyzzyzus floridanus
the last author position in a high-impact consortium publication.

Index by Category

Accident

cleanbench: *Orthotrichia garbunga*

microscopy: *Chrysotrichia likliklang*

noisy parts in a homogeniser: *Grindelia oolepis*

sudden realisation of: *Oopsacas minuta*

titration: *Boophilus microplus*

unbalanced centrifuge: *Pseudomalaxis centrifuga*

Alarm

false: *Asio flammeus sandwichensis*

Answer

invariant: *Manica yessensis*

wrong: *Idotea whymperi*

Attitude

noble but stupid: *Lorius lory jobensis*

unnecessary display of: *Dicranella stickinensis*

Autoclave

colour transmutation by: *Limonia yellowstonensis*

shape shifting by: *Gomphus vastus*

Bioinformatics

advantage: *Homalota flexibilis*

Buffer

blocking: *Laminoppia blocki*

living: *Cataulacus pompom*

Centrifuge

braking time: *Lophodionon calori*

noisy: *Surnia ulula ulula*

unbalanced rotor: *Pseudomalaxis centrifuga*

Chromatography

amount of sample lost in: *Parathemisto oblivia*

gravity flow annoyance: *Pheidole ridicula*

Cloning

bad luck: *Saccodon terminalis*

colony less: *Monodonta perplexa*

colony more: *Monodonta confusa*

obstacle to: *Mangelia acloneta acloneta*

success in: *Tachina eurekana*

undead cloning vector: *Zombia antillarum*

Coffee

marking used documents with: *Stainforthia concava*

stock solutions: *Nesaecrepida asphaltina*

visitors dropping by for a cup of: *Caffrowithius planicola*

Cold room

indicator for summer: *Zenopsis nebulosa*

Collaboration

unequal: *Busycotypus plagosus*

Communication

advice: *Ceratobaeus toheedi*

argument about travelling expenses: *Hardyadrama presignis*

Index by Category

auditory problems in:
Abudefduf luridus

basic problem in: *Manica yessensis*

coordination and timing:
Athene noctua impasta

important personal question:
Udotea javensis

lost in translation:
Tomosvaryella translata

non-verbal: *Janirella ornata*

on the way out: *Joryma sawayah*

self restraint: *Flabellum lowkeyesi*

signalling for after work beer:
Gulo gulo gulo

unscheduled: *Hya minuta*

wrong answer: *Idotea whymperi*

Computer

network problems:
Netrium digitus constrictum

wasting time with online games:
Hylaeus tetris

Congress

in exotic places:
Kanaloa kahoolawensis

smokers: *Conwentzia psociformis*

Control

handy: *Simulium duplex*

Creativity

welcome invitation to:
Wockia asperipunctella

Credibility

of lab pictures in the press:
Pseudomyrmex pictus castus

Curse

contracted by ignorance:
Tutankhamen pteromerus

effects of a: *Trefusia magna*

Daily grind

bureaucracy: *Mandragora officinarum*

Dance

experimental: *Cloeodes waltzi*

Device

puzzling: *Zoila eludens*

unused: *Usnea vainioi*

Dictionary

useful as wedge: *Ninox boobook*

Diet

unchanged: *Pipiza quadrimaculata*

Discovery

of being too late: *Anticlimax athleenae*

overrated: *Allodape greatheadi*

Distraction

welcome: *Hylaeus tetris*

Documents

cacographic notes in: *Nothrus discifer*

marking: *Stainforthia concava*

Draft

persistent: *Sweltsa continua*

Dual use

anvil for bending spatulae:
Biton adamanteus adamanteus

dictionary: *Ninox boobook*

labware: *Cora dualis*

Index by Category

manuscript: *Apolysis zzyxensis*

surprise extra feature:
Cryptocephalus obsoletus

Electrophoresis

leaky castings: *Puncturella agger*

leaky chamber: *Gelidium irregulare*

spending time at the transilluminator:
Rubus spectabilis

tool: *Scrapter vicinger*

wrong polarisation: *Hypostomus pagei*

Electroporation

excess salt in an: *Zappa confluentus*

End

of everything: *Oligia exhausta*

of late night experiments:
Nomada babai

of microscope resolution:
Microspio granulata

of the year spending frenzy:
Asellus quicki

Equipment

battered: *Abacidus fallax*

impractical: *Bembidion transparens*

improvised repair of:
Camponotus substitutus multiplis

permanent markers:
Bunchosia polystachia

missing parts:
Camponotus reticulatus fullaway

on TV shows: *Paraliparis csiroi*

other uses for: *Cora dualis*

petrification of plasticware:
Porcellio transmutatus

prehistoric megascience:
Megaceryle maxima gigantea

ignorance towards: *Ramalina obtusata*

ringtones: *Pinguicula vulgaris*

to vent aggression: *Crangon handi*

unplugged: *Disconectes ovalis*

Essay

antibody: *Sinocyclocheilus halfibindus*

Exaggeration

of achievements: *Allodape greatheadi*

Experiment

bruises by excessive handling:
Calomys sorellus

careful arrangement of:
Rhinogobius linshuiensis

delayed by shipment: *Kuwaita magna*

evasive: *Schedonnardus paniculatus*

extended duration of:
Enallagma doubledayi

important resource for:
Canbya candida

interruption:
Asio flammeus domingensis

irritation during: *Pulex irritans*

less than perfect timing: *Yoldia myalis*

messed up titration:
Boophilus microplus

most important: *Cavaticovelia aaa*

perfect timing: *Yoldia secunda*

real duration of: *Trixagus trivialis*

side effect: *Cancer productus*

social: *Deinococcus radiodurans*

unknowing: *Concoctio concenta*

Index by Category

unpredictable course of:
Cloeodes waltzi

variable outcome of: *Baris futilis*

Feng shui

in western transfer:
Rhinogobius linshuiensis

Freezer

archiving samples with: *Zapada glacier*

extra force after closing: *Stenus vacuus*

searching the -80: *Vostox excavatus*

Fun

prank: *Aha ha*

using the phrase "told you so":
Kickxia elatine

with dry ice: *Atheta surgens*

Function

extra feature:
Cryptocephalus obsoletus

Funds

amount approved: *Ecnomus dispar*

excess: *Asellus quicki*

Gel destruction

by comb: *Gelidium decompositum*

by special tool: *Scrapter vicinger*

Global warming

localised: *Notidobia demelti*

Glycerol stock

dead: *Quinqueloculina arctica*

Goo

ancient sample: *Viscosia dubiosa*

by autoclave: *Gomphus vastus*

by microwave: *Upupa epops epops*

living: *Cataulacus pompom*

on scalpel: *Scalpellum gibberum*

stuck to the centrifuge wall:
Trichoplax adhaerens

Happiness

of having achieved something:
Opius differentiarius

finishing early: *Octopus minor typicus*

Hazard

by improved safety measures:
Nosodendron slipinskii obtectum

by safety measures:
Nosodendron slipinskii

High throughput

deity of: *Rama chandramara*

Idea

lack thereof:
Elapsoidea semiannulata moebiusi

unexpected: *Ingenia mirabilis*

vague: *Dysidea amblia*

Illness

caused by missed deadlines:
Oarisma garita

Incubator

loose parts in an: *Turnix tanki tanki*

Ingredient

dirt: *Ampulex crudelis*

undisclosed blocking buffer:
Laminoppia blocki

Index by Category

Kits

dependency on: *Qianlabeo striatus*

missing components:
Parthenicus deleticus

surplus components:
Parthenicus obsoletus

Lab animal

tragic escape:
Iridomyrmex chasei yaglooensis

side effect of inbreeding:
Immergentia suecica

Lab cleanliness

areas missed underneath:
Cleantioides rotundata

centrifuge: *Trichoplax adhaerens*

scalpel: *Scalpellum gibberum*

Lab meetings

anticipation of the end: *Gulo gulo gulo*

Lab seminar

nasty colleagues: *Circus buffoni*

schedule: *Circus approximans*

Label

ambiguous: *Martiodrilus heterostickon*

cold sensitive adhesive:
Quadroppia obsoleta

deceptive: *Accipiter imitator*

Lecture

background noise: *Baobabula impolita*

lack of questions: *Anisonema pasilence*

laser pointers: *Petalium alaseriatum*

monologue in disguise:
Notoscopelus elongatus elongatus

safety: *Boreus borealis*

unintelligible: *Abudefduf luridus*

wake up call: *Wamokia remota*

Legacy

electronics: *Abacidus fallax*

passed on samples: *Viscosia dubiosa*

petrified labware:
Porcellio transmutatus

Makeshift

ashtray: *Sigara macropala*

float: *Bathylaimella simplex*

Meetings

disguise: *Pseudomugil inconspicuus*

red herring: *Pericompus elegantulus*

too many badges from:
Melitta montana

Memory

enzyme addition: *Mixojapyx saussurei*

keeping track of dispensing:
Rhipidura drownei

reminders: *Nemorimyza posticata*

sample storage: *Tetraponera demens*

Mental health

anger management: *Heliangelus viola*

discoverers Tourette's:
Thuretella schousboei

electroporation anxiety:
Megachile bangana

paranoia:
Schizopera clandestina clandestina

sequence dyslexia: *Pelexia adnata*

Index by Category

Microwave

incompatible with:
Upupa epops epops

Noise

breaking glass slide:
Chrysotrichia likliklang

centrifuge: *Surnia ulula ulula*

during a lecture: *Baobabula impolita*

long term effects:
Deinococcus radiodurans

phone calls in a lecture:
Wamokia remota

vortexing: *Ara ararauna*

Odours

embarrassing: *Fartulum occidentale*

from sink: *Yucca constricta*

Passing time

by computer games: *Hylaeus tetris*

general: *Hylaeus distractus*

Patent

for the CV: *Juncus patens*

Person

all talk: *Roncus dragobete*

annoying: *Goodeyus goodeyi*

demanding: *Busycotypus plagosus*

diehard: *Chrysops fulvaster*

experimental nitpicker:
Quercus X exacta

jumping to conclusions:
Notiophilus obscurus

pedigree: *Canis lupus familiaris*

recycling: *Danio rerio*

reliable: *Caffrowithius planicola*

setting off a numerical time bomb:
Pachybrachis chaoticus

sponger: *Matus leechi*

tidiness-challenged:
Messor sculpturatus

travelling funds: *Kanaloa manoa*

unnerving: *Bugula elongata*

unquestioned potentate:
Psychrophrynella usurpator

Pipet

multichannel problems:
Tipula desertorum

Plumbing

problems with: *Yucca constricta*

Poster

recycling: *Epinephelus posteli*

Psychology

containment of important results:
Flabellum lowkeyesi

dead giveaway:
Melanoplus devastator conspicuus

involuntary associations:
Galeodes schach

writer's block:
Elapsoidea semiannulata moebiusi

Public relation

contradicting perception:
Juncus oxymeris

set up press photographs:
Pseudomyrmex pictus castus

Index by Category

Publication

access to: *Costora luxata*

biased: *Convolvulus scammonia*

general authorship:
Zelotomys woosnami

missing figure: *Betta picta*

multiply rejected: *Downingia insignis*

oblivious authorship:
Stauroneis ignorata

rejected: *Apolysis zzyzxensis*

second authorship: *Alosa mediocris*

sponsored: *Campanula bononiensis*

ultimate authorship:
Zyzzyzus floridanus

Questions

caught off guard at parties:
Platygaster flabellata

Important: *Udotea javensis*

lack of: *Anisonema pasilence*

people disguising their monologue:
Notoscopelus elongatus elongatus

person asking too many:
Bugula elongata

Rare

office clean sweep:
Pelmatops fukiensis

perfect quadruplicate results:
Crematogaster bingo

prodigious discovery:
Titaniloricus inexpectatovus

Real life

non-magnetic magnetic particles:
Isoperla gravitans

returning from holidays:
Euphyllia paraencora

surprise questions:
Platygaster flabellata

welcome to: *Crinia nimbus*

Results

controls in retrospect:
Esperiopsis fucorum

delayed by shipment:
Catapastus simplex

different people same experiment:
Craponius inaequalis

from breaking the smoking ban:
Nicotiana forgetiana

highly reproducible:
Crematogaster bingo

incomplete data: *Fallacia omissa*

intriguing religious artefacts:
Messor luridus virginalis

lucky: *Titaniloricus inexpectatovus*

of late night experiments:
Nomada babai

repetitive paying to the mister 10%:
Tetramorium minusculum

tiny indication against:
Indicator variegatus

tiny indication in favour of a theory:
Indicator minor minor

under duress: *Oxyethira tamperensis*

unreproducible: *Baris futilis*

warped: *Gelidium irregulare*

Sample

adopted shape by electrostatics:
Ptinus fur

Index by Category

Index by Category

Theory

makeshift: *Thiara indefinita*

pet: *Brisaster fragilis*

Thesis

parts left out of: *Periploma papyratium*

too long: *Textrema hopkinsi*

Too late

correction: *Undosa undulata*

discovery: *Anticlimax athleenae*

electrophoresis: *Hypostomus pagei*

jumping the bandwagon:
Hypania invalida

online submission:
Cicindela latesignata obliviosa

realisation of a mistake:
Oopsacas minuta

Utterance

compulsive: *Thuretella schousboei*

of awe: *Wyeomyia caracula*

Vocabulary

difficult: *Archaeopone kzylzharica*

made up:
Evoxymetopon macrophthalmus

wrong place: *Blera confusa*

Wasted

Electroporation: *Zappa confluentus*

epic experiment:
Tapinoma ramulorum inrectum

microscope slide:
Chrysotrichia likliklang

money: *Usnea vainioi*

next day in the lab:
Chrysops fulvaster

pages: *Textrema hopkinsi*

patent: *Juncus patens*

sample: *Tetramorium minusculum*

space: *Pandora inflata*

thin section: *Microdillus peeli*

time: *Hylaeus distractus*

Weekend

indication: *Viverricula indica deserti*

Worst case scenario

collateral damage:
Tapinoma ramulorum inrectum

loss of primary data: *Ulosa digitata*

utterance witnessing a:
Wyeomyia caracula